FARM ANIMAL H

D0519789

A PRACTICAL GUIDE

Other Pergamon publications of related interest

GORDON
Controlled breeding in farm animals

ANDERSON and EDNEY
Practical animal handling

KENT
Technology of cereals, 3rd edition

HILL
An introduction to economics for students of agriculture, 2nd edition

DILLON and ANDERSON
The analysis of response in crop and livestock production, 3rd edition

SHIPPEN, ELLIN and CLOVER
Basic farm machinery, 3rd edition

FARM ANIMAL HEALTH

A PRACTICAL GUIDE

by

PATRICK T. CULLEN

Cheshire College of Agriculture

PERGAMON PRESS

Member of Maxwell Macmillan Pergamon Publishing Corporation

OXFORD · NEW YORK · BEIJING · FRANKFURT

SÃO PAULO · SYDNEY · TOKYO · TORONTO

U.K.	Pergamon Press plc, Headington Hill Hall, Oxford OX3 0BW, England
U.S.A.	Pergamon Press, Inc., Maxwell House, Fairview Park, Elmsford, New York 10523, U.S.A.
PEOPLE'S REPUBLIC OF CHINA	Pergamon Press, Room 4037, Qianmen Hotel, Beijing, People's Republic of China
FEDERAL REPUBLIC OF GERMANY	Pergamon Press GmbH, Hammerweg 6, D-6242 Kronberg, Federal Republic of Germany
BRAZIL	Pergamon Editora Ltda, Rua Eça de Queiros, 346, CEP 04011, Paraiso, São Paulo, Brazil
AUSTRALIA	Pergamon Press Australia Pty Ltd., P.O. Box 544, Potts Point, N.S.W. 2011, Australia
JAPAN	Pergamon Press, 5th Floor, Matsuoka Central Building, 1-7-1 Nishishinjuku, Shinjuku-ku, Tokyo 160, Japan
CANADA	Pergamon Press Canada Ltd., Suite No. 271, 253 College Street, Toronto, Ontario, Canada M5T 1R5

Copyright © 1991 Pergamon Press plc

All Rights Reserved. No part of this publication may be reproduced, stored in a retrieval system or transmitted in any form or by any means: electronic, electrostatic, magnetic tape, mechanical, photocopying, recording or otherwise, without permission in writing from the publishers.

First edition 1991

Library of Congress Cataloging in Publication Data
Cullen, Patrick T.
Farm animal health: a practical guide/Patrick T. Cullen.
p. cm.
Includes bibliographical references and index.
1. Veterinary medicine. 2. Livestock—Diseases.
3. Veterinary physiology. 4. Animal health. I. Title.
SF793.C85 1991 636.089—dc20 90-47457

British Library Cataloguing in Publication Data
Cullen, Patrick T.
Farm animal health.
1. Livestock. Veterinary aspects
I. Title
636.089

ISBN 0-08-037500-6 Hard cover
ISBN 0-08-037499-9 Flexicover

Printed in Great Britain by B.P.C.C. Wheatons Ltd, Exeter

Contents

Preface

In April 1989 I was approached by Pergamon Press who asked me if I would update various sections of the book entitled *Health and Disease in Farm Animals* by W. H. Parker. As the work progressed it became clear that because of the huge advances in animal husbandry and disease prevention in recent years, a new book was necessary. However, I have kept the same basic philosophy as Mr Parker's work. Consequently, the aim of this new book is to give students, farmers and anyone interested in agriculture an insight into all aspects of animal health.

As economic margins for livestock enterprises become more difficult to achieve, it is essential that stockpeople are able to practise sound husbandry techniques in order to prevent disaease. To do this effectively the stockperson must have a knowledge of the animal's body, how diseases are caused and what preventive measures can be taken against specific diseases. I hope that this book goes some way to supplementing the stockperson's sound practical experience with some of the theoretical knowledge required.

The first part of the book deals with the animal's body. The section is intended to examine the internal make-up of the animal to give a better understanding of how diseases affect the various organs.

The second part deals with specific diseases and disease prevention. Whilst most diseases are included within the various chapters, some important diseases, for example mastitis, have a chapter of their own as do some topics, for example animal welfare. Throughout both sections I have tried to avoid 'jargon', but inevitably some scientific terms are essential.

The book deals with the three main species of farm livestock, namely cattle, pigs and sheep. Poultry is not included owing to the very specialist nature of the subject.

I would recommend any reader who needs more detail on particular topics to consult the Further Reading list.

Nantwich, April 1990 PATRICK T. CULLEN

Acknowledgements

My thanks go to colleagues at the Cheshire College of Agriculture for their help in producing this book including: P. R. Dale, C. K. Bishop, P. D. Green, T. S. Reeves and A. G. Blackburn.

Their comments and advice on the presentation, text and photographs proved to be most helpful, but it must be said that any errors are solely mine. Thanks are due to Caroline Mansfield for her expert typing of the various drafts and to P. Blackburn for her proof reading.

For their help and advice with the photographs my thanks are due to A. R. Burch, B. Kenworthy and N. Ryder.

I should also like to thank: Milk Marketing Board for Figs. 8.2, 8.7; Cheshire College of Agriculture for Figs. 8.9, 11.1, 13.2, 13.3, 15.3, 15.8, 17.1; Kenworthy Photographic for Figs. 8.8, 19.1, 19.3, 23.3; Fullwood and Bland for Fig. 16.1; Royal Agricultural Society of England (Pig Unit) for Fig. 14.1; Mr V. R. Vernon for Fig. 13.1; Mr P. Adams for Fig. 13.7.

Finally, I must thank my wife Cheryl and my children Michael and Joanne for their support and encouragement during many hours spent on the preparation of this book.

Part 1. Animal Physiology

Chapter 1

The skeleton

The animal's skeleton is living tissue. It has nerves and blood vessels, it can be damaged by disease, can break and repair and can adjust to changes in stress.

About two-thirds of its weight are made up of inorganic compounds, mainly calcium and phosphorus salts. The remaining third consists of a framework of organic matter, mainly fibrous material and cells. The calcium and phosphorus salts are deposited onto this fibrous organic framework to give the strength and rigidity that are required to carry out its functions.

Functions of the skeleton

To give shape to the body.
To provide anchor points for muscles and therefore movement.
As a reserve for minerals, mainly calcium, phosphorus and magnesium.
For the production of red blood cells in the marrow of some long bones.
For the protection of vital organs, for example the rib cage protects the heart and lungs.

Composition of the skeleton

The skeleton consists mainly of bone. Areas of cartilage are also present.

Bone

Bone is made up of inorganic mineral substances, mainly calcium, phosphorus and magnesium compounds, although these minerals are not permanently fixed. On the outside bone is hard and compact, but inside there is often a cavity, the medullary cavity, containing marrow (especially in long bones like those that form the limbs). Near the ends there is a type of bone known as 'spongy bone', although it is actually quite hard. Marrow is a soft tissue richly supplied with blood and is involved in the formation of blood cells (see Chapter 5).

The outside of the bone is covered by a thin membrane, the periosteum, which is supplied with blood vessels and nerves.

Growth and repair of bone

In the young animal there is a layer of cartilage within the bone that acts as a growing point. The bone will lengthen and thicken. This process is brought about by two specialist cells:

Osteoclasts—which break down the cartilage
Osteoblasts—which deposit minerals on to bone to build a rigid material.

These two cells are also involved in the repair of bone; osteoclasts dissolve any splinters of bone and osteoblasts build up new bone.

The speed of any repair depends to a large extent on the blood supply to the bone. As the bone of young animals has a better blood supply than the bone of older animals, breaks in young bone will normally heal more quickly. The broken ends must be touching and immobilized for fibrous tissues to form in the break and so start the process of repair.

A farmer who has an animal with a broken limb should consult a veterinary surgeon to find out the chances of the break repairing permanently. This will depend on the severity of the break and the age of the animal. The value of the particular animal will also be taken into account. With most commercial stock it may be best to cull the animal as soon as the damage has occurred in order to avoid unnecessary suffering.

Cartilage

Cartilage is a flexible, elastic, glossy-looking substance composed mainly of a protein called chondrin. It is present in joints to help with movement and is also found in noses, ears and the lower end of rib bones. The skeleton of a fetus is made up almost entirely of cartilage, but as the animal ages and develops the cartilage slowly changes to bone, a process known as ossification.

The animal skeleton

The skeleton of farm animals consists of four limbs, a rib cage and skull, all attached to the spine or vertebral column (Fig. 1.1).

The spine is made up of a number of individual bones, the vertebrae, which have a small degree of movement between each. However, if several vertebrae move together the sum total of the movement can be quite large, allowing the animal to get its head right round to lick and groom its flanks.

In one area of the vertebral column, the sacrum (Fig. 1.1), five individual vertebrae are fused together as one and joined to the pelvis. This extra strength is necessary because the muscles of the hind limbs, the most powerful muscles of the body, are attached to this region.

The ribs are in pairs and cover both sides of the thorax or chest cavity. The upper portion of a rib is bone and is attached to the spine; the lower end is cartilage, which, being more flexible than bone, increases the mobility of the chest wall.

Joints

Joints allow animals to move freely. Within the body there are several different structures of joint and several methods of classifying joints.

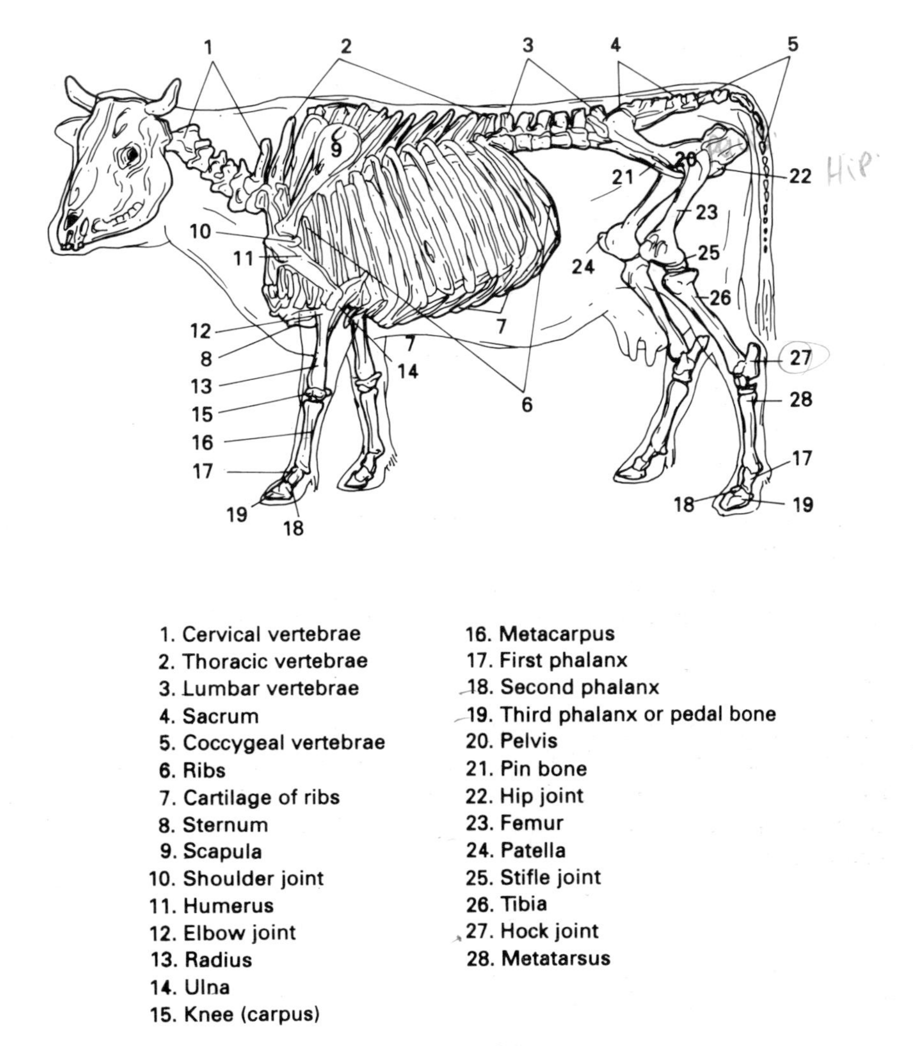

FIG. 1.1 The skeleton of the cow.

One of the simplest methods describes the degree of movement in the joint.

1. *Immovable*—The bones are in direct contact, held together by fibrous tissue which ossifies as the animal ages, e.g. the skull and a section of the spine above the pelvis—the sacrum;
2. *Slightly movable*—Small amount of movement between each joint, e.g. the joints between the individual bones of the spinal column.
3. *Movable*—Large amount of movement allowed, e.g. a hinge joint (knee) or ball and socket joint (hip).

Two basic joint structures are:

Cartilaginous joints—bones are united by cartilage, e.g. the joints between individual vertebrae.

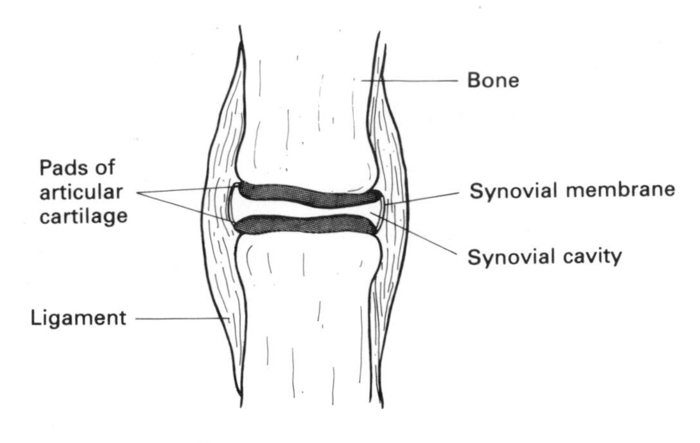

FIG. 1.2 A synovial joint

Synovial joints (Fig. 1.2)—there is a small gap, the synovial cavity, between the two bones, e.g. the elbow joint.

The characteristics of a synovial joint (Fig. 1.2) are as follows.

The articular cartilage covers the end of the bones to ease movement.
The synovial cavity is a potential cavity between the two bones. In practice the bones are almost touching, similar to a piston within a cylinder.
The synovial membrane surrounds the joint and secretes synovial fluid (joint oil) to lubricate the joint.
The ligaments are bands of connective tissue which bind the joint together.

(*N.B.* Tendons are connective tissue bands that connect muscle to bone.)

Joints in farm animals are not without their problems. The cartilage may get damaged, leading to stiffness, lameness and possible arthritis; joints can become dislocated and they can get infected with bacteria, again resulting in lameness.

The limbs

The limbs of farm animals are very vulnerable to injury and infection, so it is important that the stockperson has a knowledge of their anatomy. Fig. 1.3 shows the individual bones in the fore limb and the hind limb of a cow. The bones in the limbs
of sheep are very similar to those of the cow, whilst in the pig there is a slight variation.

The fore limb (Fig. 1.3)

1. Scapula. Farm animals have no collar bone on which to 'hang' the fore limb. Instead the scapula is held in place by being embedded in a band of very tough shoulder muscle.
2. Radius and Ulna. In humans these are two separate bones (in the forearm) to enable twisting at the wrist. In farm animals the radius and ulna are bound

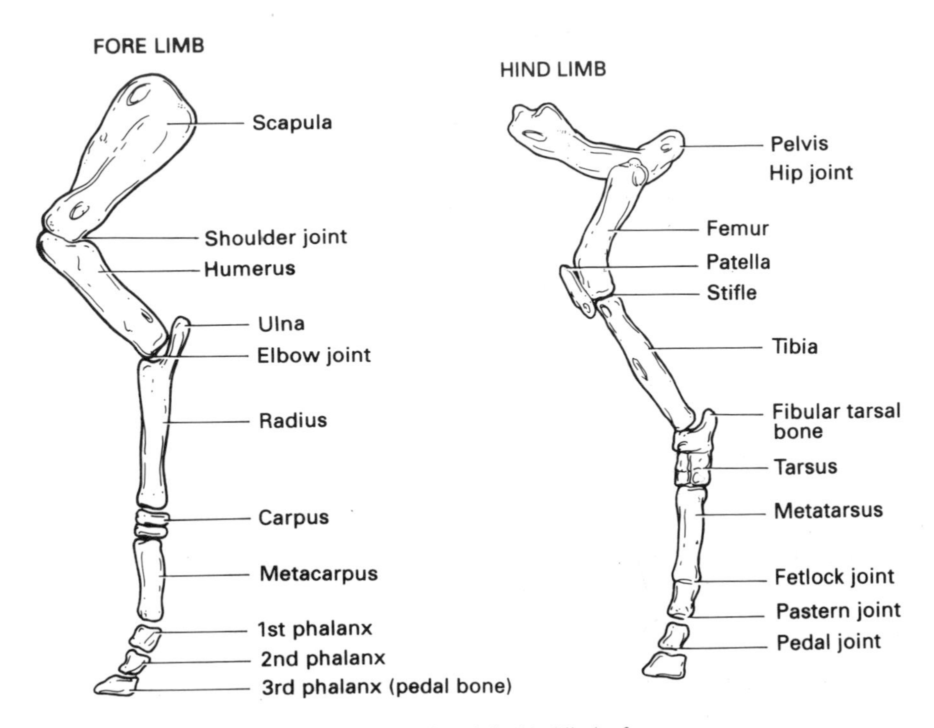

FIG. 1.3 The fore limb and the hind limb of a cow.

together by fibrous tissue, so no twisting movement is possible. The top of the ulna points up to form the point of the elbow. This provides attachment for muscles and better leverage.

3. Metacarpus. In cows this looks like one bone, except at its lower end where it divides to form the start of the cow's two-part foot.

The hind limb (Fig. 1.3)

1. Pelvis. Three bones fused together form the pelvic girdle. It is through this ring of bone that the young animal must pass at birth. In the case of the sow there are usually no problems as the piglets are so small, but problems can arise for the ewe and the cow. A common misconception is that the pelvic 'bones slacken for calving'. This is not so; it is the slackening of ligaments in the pelvis that gives the feeling of slackness.

   It is worth noting that if excessive force is used when assisting at a birth (especially calving) the bones in the pelvic girdle may 'split' and cause damage that is almost impossible to repair—the animal will normally have to be culled.
2. Femur. One of the biggest bones in the body, heavily muscled, running downward and forward.
3. Patella. The equivalent of the kneecap in man.
4. Fibular tarsal bone. This projects upwards and backwards to form a lever for attachment of the Achilles tendon.
5. Metatarsus. As the metacarpus of the fore limb.

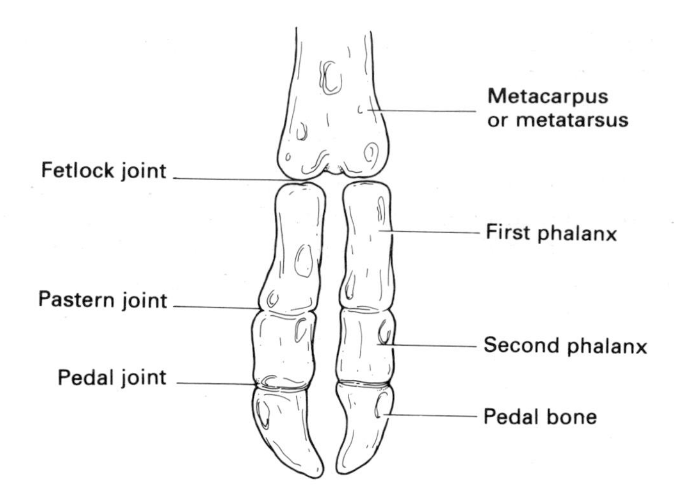

FIG. 1.4 The two-part foot of the cow.

Foot/Hoof

Below the metacarpus (fore limb) and metatarsus (hind limb), all four feet have the same structure.

In the cow's two-part foot (Fig. 1.4), each part is known as a digit and each digit is composed of three bones known as phalanges (singular: phalanx). The third phalanx, which has the same shape as the outer hoof, is commonly referred to as the pedal bone.

In Fig. 1.5 it can be seen that the outer hoof (horn) comes up to almost the centre of the second phalanx, so little movement is allowed in the pedal joint. The area where the outer wall joins the skin of the leg is known as the coronet.

The navicular bone is a small bone situated at the rear of the pedal bone.

The pedal bone is covered with a substance known as the laminae or 'quick'. The laminae is very sensitive and is richly supplied with blood vessels. It has specialist cells which produce the hard outer wall of the hoof.

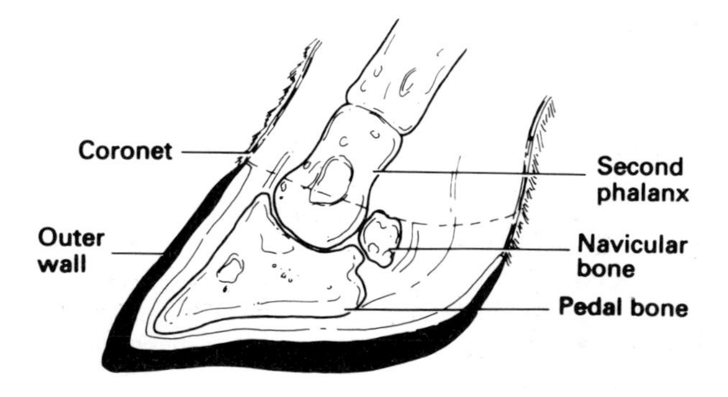

FIG. 1.5 Cross-section through a cow's foot.

The reader should note that Chapter 15 deals with lameness in cattle and covers many diseases and problems of cows' feet.

Teeth

Strictly speaking the teeth are not part of the skeletal system, as their structure and composition are different from those of bone. However, I will cover teeth in this section as stockpeople normally associate the teeth with the skeleton.

Farm animals, like humans, have two sets of teeth, temporary or 'milk' teeth and permanent teeth that erupt in later life. There are four types of teeth:

Incisors—front teeth used for cutting and biting.
Canine—commonly called 'eye' teeth, used for tearing and ripping flesh.
Premolars } large flat teeth at side and back of mouth used for grinding and
Molars } chewing.

All four types of teeth have the same basic structure (Fig. 1.6), although the size and shape do vary between the different species.

As the number of teeth varies between species, a dental formula is used to indicate the number of the various teeth to be found on each side of the mouth for any species. The following example is for the permanent teeth of cattle:

	Incisors	Canines	Premolars	Molars
Upper jaw	0	0	3	3
Lower jaw	4	0	3	3

As the formula shows the teeth in one side (or half) of the mouth only, the total number of teeth in the permanent set is found by adding all the teeth together and doubling. So the total number of teeth for adult cattle is 32.

Adult sheep have the same number of teeth as adult cattle and also have the same dental formula. Pigs, on the other hand, are not herbivores as are cattle and sheep, so they have canine teeth for tearing meat and top and bottom sets of incisor teeth.

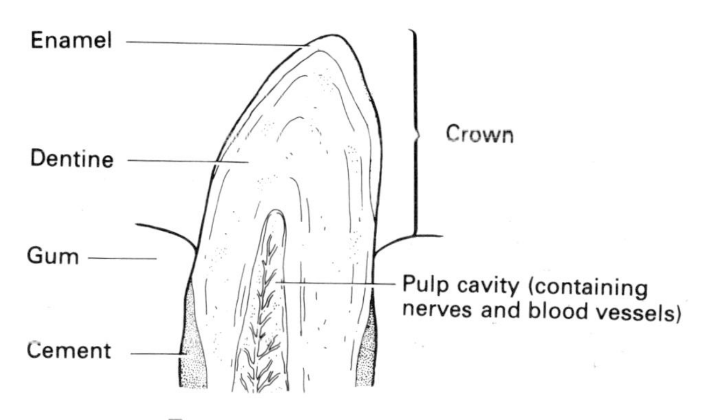

FIG. 1.6 The structure of a tooth.

The dental formula for a full set of permanent teeth in a pig is:

	Incisors	Canines	Premolars	Molars
Upper jaw	3	1	4	3
Lower jaw	3	1	4	3
		Total 44		

Ruminants have a hard dental pad in place of the upper incisors. The lower incisors tear the grass and silage by gripping it between the lower incisors and the dental pad. The gap left by the absence of the canine teeth is of considerable use when opening the mouth for examination and drenching.

Table 1.1 shows the number of temporary and permanent teeth for cattle, sheep and pigs.

TABLE 1.1 *Numbers of temporary and permanent teeth.*

	Temporary	Permanent
Cattle	20	32
Sheep	20	32
Pig	32	44

Ageing by examination of the teeth (cattle and sheep)

Estimation of age by this method is not an exact science. The best way of finding out the age of an animal is to look up the records. However, this system can be used if you need a rough guide and can be especially useful when buying sheep.

The method involves the use of the incisor teeth only, usually referred to as broad teeth. The central pair (one on either side of the mid-line of the mouth) of temporary teeth are replaced first, then the next pair and so one (Fig. 1.7) until all the

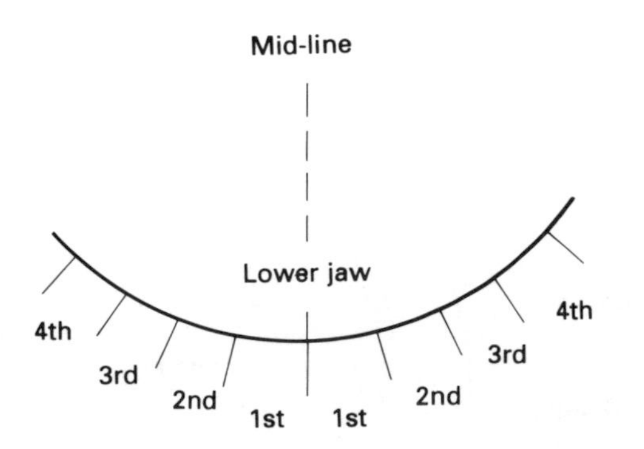

FIG. 1.7 Diagram showing the order of eruption of pairs of broad permanent teeth in cattle and sheep.

temporary teeth have been replaced by eight broad (permanent) teeth. The replacement follows this approximate time scale:

Cattle:

1 year 6 months—2 broad teeth
2 years 3 months to 2 years 6 months—4 broad teeth
2 years 9 months to 3 years—6 broad teeth
3 years 9 months to 4 years—8 broad teeth

Heifers on a self-feed silage system may have problems as their teeth change from temporary to permanent, especially if the clamp is very compacted. They may be unable to pull out sufficient silage for their needs, so a close watch must be kept on their body condition.

Sheep:

1 year 3 months—2 broad teeth
1 year 6 months to 1 year 9 months—4 broad teeth
2 years 3 months to 2 years 6 months—6 broad teeth
2 years 9 months to 3 years—8 broad teeth

As sheep get older their permanent teeth become loose and are eventually lost. Such an animal is referred to as 'broken mouthed'. Diets where the animals have to work for their food, for example root crops or coarse upland grazing, are not suitable for older ewes. Their productive life may be lengthened by a few years if they are moved on to kinder diet, a lowland pasture for example.

Pigs' teeth are more erratic in their eruption and so are not used to assess their age.

Piglets are born with very sharp 'eye' teeth that can cause soreness to the sow's udder and may give their litter mates a nasty scratch if they decide to fight. It is the practice on some units to clip these teeth off as soon as the piglet is born.

The canine teeth or tusks of a boar can grow to 6–9 cm in length. As these are a potential source of danger to the stockperson, they may be removed by a veterinary surgeon.

Chapter 2

The muscular system

The selling of muscle (meat) is a major part of agriculture, indeed the beef, sheep and pig industries depend on it. The function of muscle is to enable the animal to move its limbs and various other parts of the body. There are three types of muscle.

Voluntary or skeletal muscle

This forms the bulk of body muscle and, therefore, of 'meat'. The typical voluntary muscle has a tendon at each end, usually fixed to a bone, with the muscle block in between. The muscle is composed of fibres grouped together in bundles (Fig. 2.1).

Each fibre is supplied with a nerve and a blood vessel. When the nerve is stimulated,

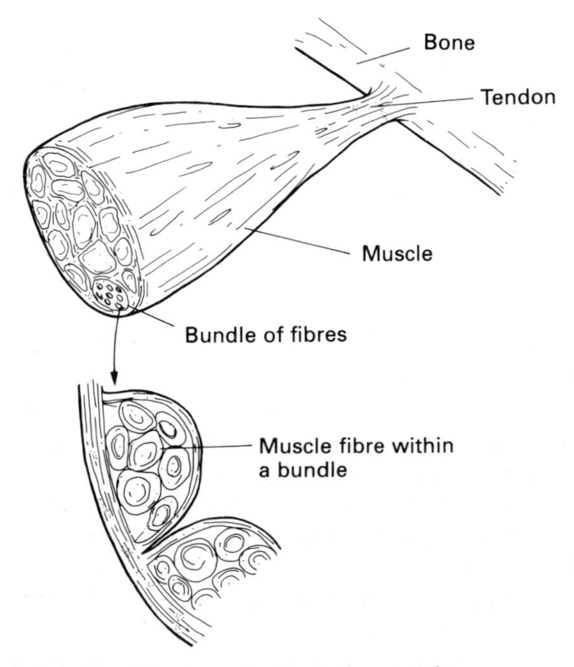

FIG. 2.1 Structure of voluntary muscle.

the fibre contracts. The stronger the nerve impulse, the more muscle fibres are stimulated and the more powerful the contraction. It is worth noting that muscles can only contract (pull) and relax, they cannot push. Therefore, most muscles work in pairs to enable a limb to move backwards and forwards.

Voluntary muscle is under the direct control of the animal's will.

Involuntary or smooth muscle

This type of muscle, which has a different structure from voluntary muscle, is found around internal organs, e.g. the gut and blood vessels.

This means that food can be moved along the gut by wave-like contractions of the muscle (peristalsis) and that blood vessels can expand and contract when necessary.

These muscles work independently of the animal's will and are controlled by the central nervous system.

Cardiac (heart) muscle

This is an involuntary muscle with a different structure to both voluntary and smooth muscle.

It has a short but very strong contraction followed by a period of rest, thus the beating action of the heart. One side of the heart contracts whilst the other rests.

It acts independently of the body, but hormones can alter the speed of the beat (notably adrenalin).

Meat quality

There are many factors that can influence the quality of the meat that the farmer sells to the slaughterhouse, ranging from the genetic make-up of the individual animal to its age and sex. However, the behaviour of the animal just before slaughter can also have a direct effect on meat quality.

The muscle has a store of energy in the form of glycogen which, when used by the muscle, releases energy in the form of movement and heat. As the glycogen gets used, it is converted into lactic acid which lowers the acidity of the muscle.

If an animal is put under stress just before it is slaughtered, its muscular activity is increased and it uses up the glycogen in the muscles. When the animal is slaughtered there is little or no lactic acid in the muscle. This raises the pH of the meat, making it more alkaline, and results in poor keeping quality.

In beef animals, especially young bulls, pre-slaughter stress results in a condition known as 'dark-cutting'. Apart from unappetizing appearance of the meat, the high pH of 'dark-cutting' beef encourages the growth of bacteria.

The connective tissue between the muscle bundles is known as collagen. Collagen is partly responsible for making meat 'tough' to eat. The collagen is broken down during the cooking process, making this meat more tender. Soon after slaughter, natural enzymes in the meat begin to break down the collagen. Therefore, meat is hung for a short while in the slaughterhouse to increase its tenderness.

Different parts of the animal contain different amounts of collagen and also different sizes of muscle bundles. This is why rump steak with fine muscle bundles and little connective tissue is more tender than a coarse piece of shin beef.

As the animal ages, the amount of collagen increases, so the meat of young animals is normally more tender than that of older ones. It is said, however, that meat from older animals has more flavour.

Chapter 3

The digestive system

One of the most important jobs on any farm is the correct feeding of the animals. In order to do this effectively the stockperson should know how each type of animal digests its food. This chapter will deal with the structure of the digestive system and not with the details of the chemicals involved in digestion or the chemical outcome of the breakdown of food.

There are two types of digestive systems in farm animals:

Simple-stomached, or the non-ruminant digestive system, e.g. pigs.
Multi-stomached, or the ruminant digestive system, e.g. cattle and sheep.

Apart from the structure, there is another important difference between the two. In non-ruminants the food is broken down mainly by chemical enzymes, whereas ruminant digestion is brought about by microorganisms and enzymes. Thus ruminants are able to utilize fibrous foods such as grass and silage.

The non-ruminant digestive system (Fig. 3.1)

The process of digestion starts in the mouth when food is chewed; this breaks it down into small pieces to increase the surface area on which the enzymes will act. Saliva, produced by three pairs of salivary glands in the mouth, lubricates the food and a chemical in the saliva begins to break down the food. The tongue forms the food into a ball (bolus) which is swallowed. It passes down the oesophagus and into the stomach. The entrance to the stomach has a ring of sphincter muscle to stop the stomach contents from going back up the oesophagus.

The stomach is a large, thick-walled muscular sac. It is capable of storing large quantities of food which get mixed and churned by muscular action. The walls of the stomach secrete mucus, hydrochloric acid and the enzymes which continue the digestion process started in the mouth.

A sphincter muscle at the exit from the stomach allows small quantities of food to pass into the intestines.

The intestines are coiled muscular tubes with a number of distinct sections (Fig. 3.1).

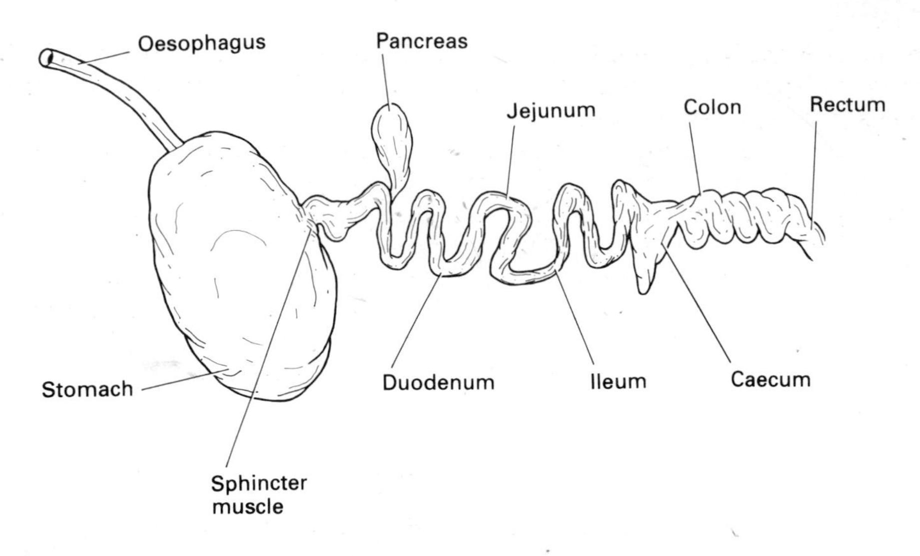

FIG. 3.1 The non-ruminant digestive system.

The small intestine consists of—Duodenum
—Jejunum
—Ileum
The large intestine consists of—Caecum
—Colon
—Rectum

Two structures secrete juices into the duodenum. The pancreas secretes pancreatic juices containing enzymes which continue the digestion process. The gall bladder, attached to the liver, secretes bile which emulsifies the fat in the diet. Bile, which is alkaline, neutralizes the acidity of the contents of the intestines.

The small intestine is where the digestion and absorption of the broken-down food takes place. The internal structure (Fig. 3.2) is designed to increase the surface area for the absorption of the food. It has finger-like projections called villi which contain

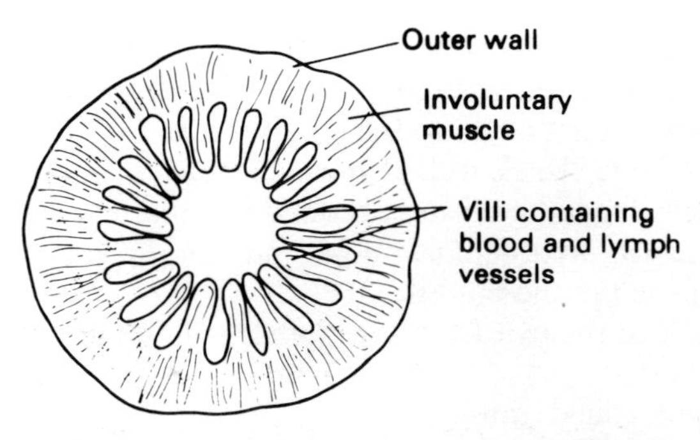

FIG. 3.2 Internal structure of the intestine, showing villi.

blood vessels and lymph vessels. The broken-down food, in the form of carbohydrates and amino acids, is absorbed into the blood system. The fats are absorbed into the lymphatic system (*see* page 28).

The large intestine begins at the caecum where a limited amount of bacterial breakdown of the fibrous parts of the food takes place. It is interesting to note that horses, which are non-ruminants, have a large caecum which enables them to utilize coarse fodder such as grass.

The colon is involved in the absorption of water from the contents of the intestine back into the body, leaving a dryish material to collect in the rectum which is voided at regular intervals. If an animal is scouring, this process of reabsorption of water is upset and the animal will dehydrate, possibly to the point of death.

Ruminant digestion

In order to break down a fibrous diet, the ruminant has evolved a specialist digestive system. It has four 'stomachs' (although only one is a true stomach) and relies on microorganisms to break down the fibre. The rest of the digestive system is the same as in the non-ruminant.

The four 'stomachs' are (Fig. 3.3):

Rumen.
Reticulum.
Omasum.
Abomasum.

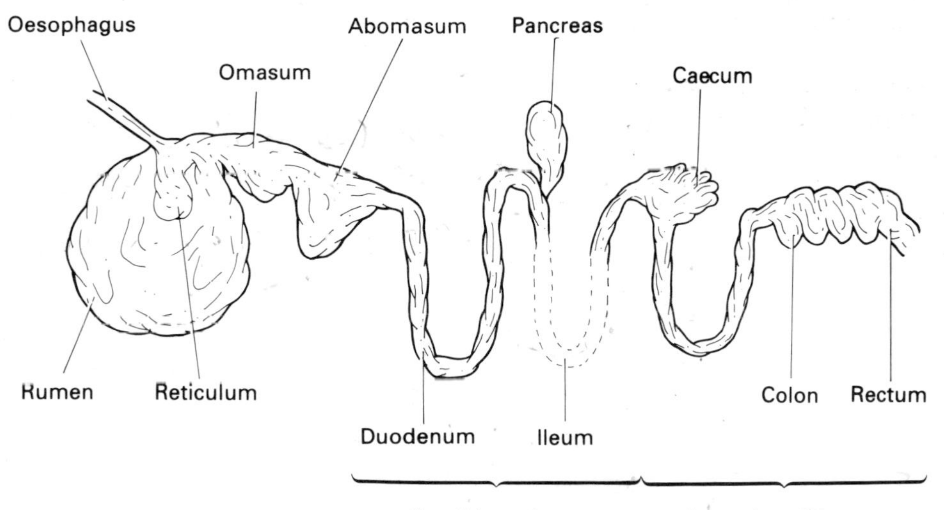

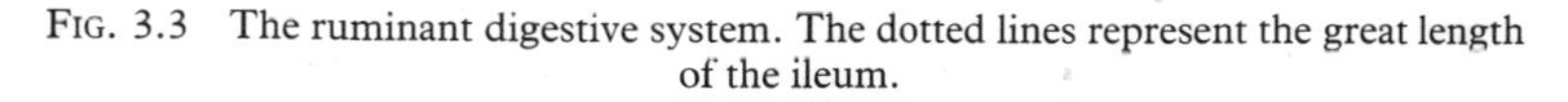

FIG. 3.3 The ruminant digestive system. The dotted lines represent the great length of the ileum.

Rumen

The rumen is the most important of the four stomachs and is the site where most of the food is broken down. It is situated on the left of the animal and takes up more than half the abdominal cavity. The walls are tough and muscular in order to contract and mix the rumen contents. The inside walls have a furry texture which increases the surface area on which bacteria and protozoa may lodge.

The salivary glands in the mouth produce large quantities of saliva; this is important in decreasing the acidity of the rumen contents, as acids are produced during the fermentation and subsequent break-down of the diet. Excess acid in the rumen can lead to a digestive problem known as acidosis (*see* page 155).

Large amounts of gas are also produced during fermentation, mainly methane and carbon dioxide. If these gases are not expelled by belching, it can lead to bloat (*see* page 155) and possibly the death of the animal.

Reticulum

This joins the front of the rumen and acts as an extension of it. Heavy objects such as nails, stones and wire that are accidentally swallowed by the animal fall into the reticulum and are trapped. As the reticulum is close to the heart, nails and wire can occasionally pierce the heart. If wire does get into the reticulum, a veterinary surgeon will have to carry out an operation to remove it.

Fibrous material let through from the rumen floats on top of the liquid and is picked up by a muscular groove which runs along the top of the rumen and reticulum—the oesophageal groove. This fibrous material is then passed back to the mouth for chewing, i.e. chewing the cud.

This system has a number of advantages. Firstly, the animal is able to take in large amounts of food and then go somewhere quiet to digest. Secondly, the very coarse food gets thoroughly ground, thus increasing the surface area for the bacteria to work on. Once chewed, the bolus of food is re-swallowed back into the rumen.

The amount of chewing an animal does depends on the type of food it is eating. A diet of coarse food leads to more chewing, a diet rich in concentrates leads to less. Chewing the cud is an important sign of health in ruminants.

The oesophageal groove is used in young calves and lambs to by-pass the rumen and take milk directly to the omasum and abomasum.

Omasum

This is a small stomach with folds of muscle hanging from the roof rather like the pages in a book. Indeed, an old name for the omasum was 'the bible'. The food gets caught between the muscular folds and is squeezed. The liquid then passes to the abomasum.

Abomasum

This is the true stomach, similar to that of the pig and with the same structure as described in the section on the non-ruminant.

The rest of the digestive tract is very similar to that of the non-ruminant, including secretions by the pancreas and the gall bladder and absorption by villi.

The complexity of the ruminant's digestive system can lead to problems. One of these is that, unlike single-stomached animals, ruminants cannot usually vomit, so anything ingested stays in the stomach and cannot be removed except surgically.

The liver

Although the liver is not directly related to the topic of digestion, it is such an important organ that it would be wrong not to make reference to it.

In ruminants the liver is situated behind the diaphragm (below the lungs) and partly overlapping the rumen on the right. It is a large organ weighing 3–4 per cent of the total body weight.

Blood from the stomach and intestines enters the liver via the hepatic portal vein. The liver is then responsible for carrying out a number of functions; the main ones are listed below.

1. The formation of bile from worn-out red blood cells. Bile is stored in the gall bladder. It is a greenish-yellow liquid used in the digestive process to emulsify fat. The gall bladder is attached to the liver, the liver having many bile-collecting ducts which drain into the gall bladder.
2. The breakdown of excess amino acids. Excess protein in the diet can lead to excess amino acids. These are broken down to release energy and urea, the urea being excreted via the urine.
3. The storage of some fat-soluble vitamins, e.g. A, D and B12.
4. The storage of iron that is recovered from the breakdown of the red blood cells.
5. The breakdown of body fat reserves. Animals that are short of energy for whatever reason, possibly poor diet or high production, will use body fat to make up the deficit. The breakdown of these fats leads to toxic ketones being released, sometimes resulting in the metabolic disorder known as acetonaemia (*see* page 148).
6. The regulation of blood sugar levels by the storage of glycogen. Some sugars that have been released by the digestive process are stored in the liver until needed. This evens out the supply of energy between feeds.
7. The removal of some toxic substances from the blood.

Chapter 4

The respiratory system

All the time, in all living animal cells, energy is being released. This energy results from the oxidation of complex carbon-containing molecules, and therefore a constant supply of oxygen is necessary to sustain life. The oxidation of glucose (a simple sugar) is roughly similar to the oxidation process that goes on within each cell and can be illustrated by the formula.

$$\text{Oxygen} + \text{glucose} \rightarrow \text{broken down into} \rightarrow \text{carbon dioxide} + \text{water} + \text{energy}$$

$$\text{i.e. } 6O_2 + C_6H_{12}O_6 \rightarrow 6CO_2 + 6H_2O + \text{energy}$$

One of the waste products of this break-down is carbon dioxide. Because this is toxic, there needs to be some means of elimination.

Respiration is the major process for the elimination of carbon dioxide as well as the principal process by which cells obtain oxygen for the continuing oxidation process.

Structure of the respiratory system

The respiratory organs (Fig. 4.1) are the only internal organs of the body that are constantly exposed to the outside atmosphere. The air passages leading to the lungs can be categorized as follows:

Nasal cavities.
Pharynx (upper throat) (not shown in Fig. 4.1).
Larynx (voice box) (not shown in Fig. 4.1).
Trachea (windpipe).
Two bronchi which lead into the lungs themselves.

The nasal cavities are lined with a mucous membrane. In the upper part of each nostril are turbinate bones which are shaped rather like scrolls of paper. Their function is to warm the air as it passes through. Small hairs trap dust particles so that warm, clean air is breathed into the lungs.

The pharynx is in the upper throat and, together with the epiglottis, is involved in sealing off the oesophagus when breathing and sealing off the trachea when swallowing food.

The larynx is divided into two cavities. A fold of mucous membrane from each side

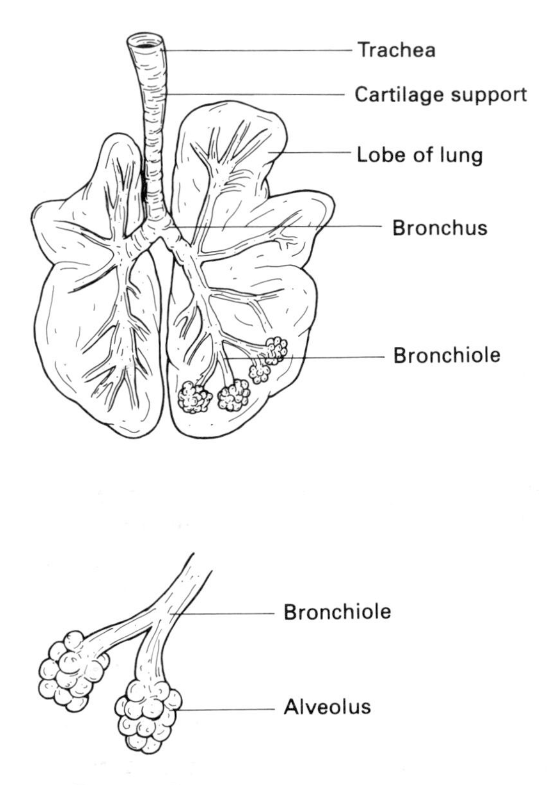

FIG. 4.1 The respiratory system of a sheep.

forms a pillar at the division. These folds are the vocal cords controlling the sound the animal makes as air passes between them.

The trachea (or windpipe) is a non-collapsible tube, supported by cartilage, that divides into two bronchi (singular: bronchus) which enter each lung. Each bronchus then divides into smaller alveolar ducts. Under the microscope the alveolar ducts can be seen to have grape-like projections only one cell thick—the alveoli. It is in these alveoli that gaseous exchange takes place.

Gaseous exchange takes place in the alveoli by diffusion through the walls of the blood vessels and the walls of the alveoli. The gases are in solution in the blood on one side and in the moisture lining the alveoli on the other. The diffusion of these dissolved gases takes place from an area of higher concentration to an area of lower concentration for each gas.

The lungs are made up of millions of alveoli so that when they are removed from the body the lungs look pink and spongy. They are divided into lobes to increase the surface area and are covered by a membrane called the pleura. If an animal contracts a respiratory disease it is often the alveoli that become infected. Once damaged they may never again function properly.

The airways of the smaller bronchi and bronchioles are cleaned by hair-like projections (cilia) on the lining cells. These have a wave-like motion which moves mucus and foreign matter up to the epiglottis where it is normally swallowed.

The mechanics of breathing

Air enters the lungs because of atmospheric pressure. Inside the chest cavity the lungs are enclosed, for all intents and purposes, in a vacuum. The diaphragm, which is a large sheet of ligament at the base of chest area, is moved down and the ribs swivel outwards increasing the volume of the chest. Air rushes into the lungs which expand to fill the increasing chest cavity. Expiration is a relaxation of the chest muscles and thus a reversal of the process (Fig. 4.2).

Rates of respiration (Table 4.1) vary between animals. Young animals usually breathe faster than adults and most species breathe faster in hot weather. When cows and sheep chew the cud the rates increase to 60–80 breaths per minute in cattle and up to 100 in sheep. Breathing rate and breathing action are points to look for when checking the health of stock. If an animal seems to be breathing faster or its chest is 'thumping' as it breathes, it could be a sign of a chest infection.

The start of breathing at birth

The fetus 'breathes' by extracting oxygen from the blood of its mother until the umbilical cord is broken. At that point oxygen starvation begins to build up rapidly,

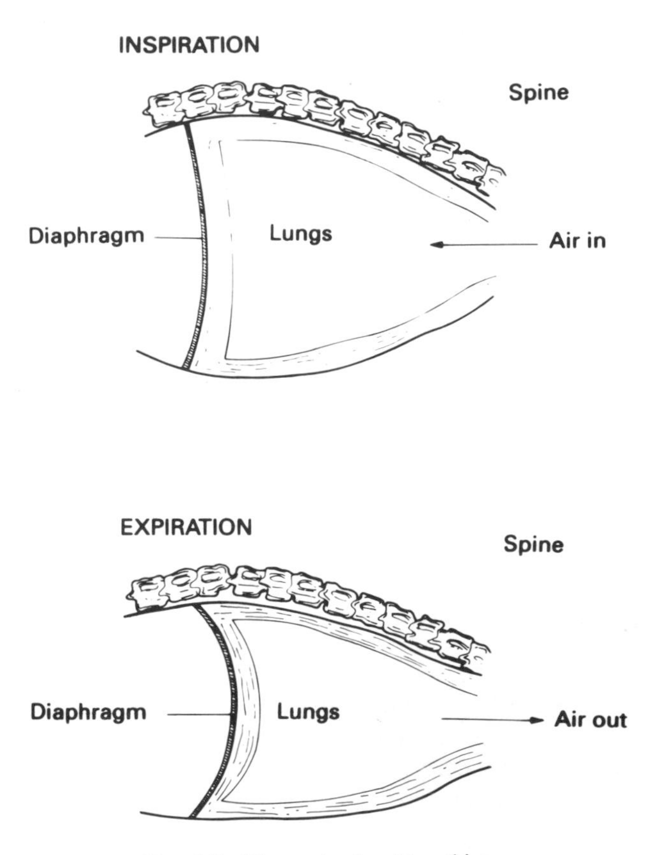

FIG. 4.2 The mechanics of breathing.

TABLE 4.1 *Respiration rates of animals—adults under cool conditions.*

Species	Inhalations per minute
Cattle	12–16
Sheep	15–20
Pig	10–20

but the accumulation of carbon dioxide stimulates the respiratory centre of the brain. Because the chest has never been used to breathe before, a powerful stimulus is needed to make it respond. It receives this stimulus when the warm, wet chest of the newborn animal experiences the cold of the outside world and certain nerve endings on the chest are stimulated, making the animal gasp and gulp in its first lungful of air.

In cases where an animal is born but fails to start breathing, one technique, to be used only as a last resort, is to pour a bucket of cold water over the animal in order to stimulate these nerves. It is these same nerves that react when we are suddenly immersed in very cold water; we involuntarily expand the chest and gulp the air. Rubbing the chest of a newborn calf or lamb can also help to initiate the breathing response.

Chapter 5

The circulatory system

The circulatory or vascular system is composed of both the blood system and the lymph system.

The blood system

Single-celled animals have no circulatory system. Their surface area is large in relation to their volume and so all parts of the organism are close to the outside atmosphere. However, as body size and, therefore, volume increases, the surface area of the body is too small and the internal body organs are too far from the outside for diffusion of gases to take place and so a transport system is required. The circulatory system links the major organs of the body in order that the body acts as one unit. It is rather like the railway system connecting the major cities of the country.

The functions of blood and the circulatory system are as follows:

To carry oxygen from the lungs to all living body cells.
To remove carbon dioxide and other waste products from body cells and take it to disposal points such as the lungs and kidneys.
To carry food to all cells of the body.
To carry hormones from endocrine glands to various parts of the body.
To move antibodies and white blood cells to sites of infection in order to prevent disease spreading throughout the body.
To help regulate body temperature.
To keep body water content stable.
To keep body pH stable at 7.4.

In order to understand the blood system we need to study three distinct items: the blood itself; the blood vessels; and the heart.

Blood

The blood of cows, pigs and sheep comprises about 8 per cent of the total body weight and is composed of 40 per cent cells and 60 per cent plasma. The plasma, a straw-coloured liquid, carries nutrients, salts and carbon dioxide in solution and so is responsible for a number of the functions of blood noted earlier. The cells can be divided into three groups with specialist functions.

Red blood cells (erythrocytes)

These are present in large numbers, about six million per cubic millimetre of cows' blood. Under the microscope the cells appear as red biconcave discs. They live for about six weeks and are broken down by the liver (*see* Chapter 3). New red blood cells are formed in bone marrow and continually replace the worn out cells.

The red colour is due to a protein called haemoglobin which contains iron. The formulation of red blood cells is severely restricted if iron is not available to an animal. Piglets are born with low reserves of iron and so are injected with iron within the first few days of life to prevent anaemia. Piglets kept outside do not need this injection as they pick up iron from rooting in the soil. The haemoglobin has the ability of readily combining with oxygen when it is available and releasing oxygen when it is scarce. So haemoglobin combines with oxygen in the lungs and carries it to the tissues of the body.

White blood cells (leucocytes)

There are not as many white blood cells as red blood cells, and the number varies between species. The average number per cubic millimetre of cows' blood is 7000–8000; in the pig there are about 17,000.

There are several different types of white blood cell, but the main function of them all is to protect the body against invasion by bacteria and other organisms. White blood cells have a life of a few days rather than weeks.

For our purposes some information on two types of white blood cells seems appropriate.

1. Lymphocytes. These are small round cells with a large nucleus and are capable of independent movement similar to 'amoeboid movement'. They are formed in the lymph nodes, the spleen, and in bone marrow. They can produce antibodies to kill bacteria.
2. Granulocytes. Again, these are capable of amoeboid movement. Their numbers increase rapidly if the animal is infected and they can migrate through the walls of capillaries to the affected areas. The severity of a number of diseases can be estimated by counting the number of granulocytes present in a sample of blood.

Platelets

There are about 200,000 platelets per cubic millimetre of blood. Their function is to clot the blood in order to avoid blood loss from a wound.

A platelet is a frail, easily damaged cell. In a healthy blood vessel the platelets remain intact, but if a breakdown occurs which causes bleeding from that blood vessel, whether it is internal bleeding or bleeding through the skin, some of the platelets rupture on the rough surface of the broken blood vessel or skin. An enzyme is then released from the platelets and sets in motion a chain reaction resulting in the formation of fibrin, a sticky mesh of material which traps large numbers of escaping blood cells. The fibrin contracts and squashes the cells together, tending to seal the wound and stop the bleeding. A cut made by a sharp blade such as a scalpel has a smooth edge. Few platelets are broken so the clot is slow to form and a lot of blood can be lost. On the other hand, a jagged cut, say from barbed wire, causes more tissue damage and more platelets are broken, resulting in a clot forming quickly with relatively little loss of blood.

Blood vessels

The vessels involved in the circulatory system are:

Arteries.
Veins.
Capillaries.

There are other important organs in the body that are closely related to the circulatory system, but these are dealt with separately, for example the lungs, the liver and the kidneys.

Arteries take blood away from the heart, veins carry blood back to the heart. There are no exceptions. Arteries have to withstand the force of the heartbeat which drives blood through them, so the wall of an artery is more elastic and muscular than the wall of a vein. As blood leaves the heart and enters the main artery there are one-way valves to prevent the blood flowing back into the heart.

There are few other valves in the arteries. However, there are a number of valves in veins taking the blood to the heart; as the force of the heartbeat is weaker in the veins, the blood needs the valves to prevent any backflow. The main artery, the aorta, leaves the heart. The arteries then divide down, the bore getting smaller until the smallest, the arterioles, can only be seen under a microscope. Veins have a similar pattern, but the blood enters the venous system by the minute venules and goes through progressively larger vessels until the main vein enters the heart.

Capillaries are the smallest blood vessels and are the link between arterioles and venules. It is here that the exchange of gases, food and waste products takes place between the blood and the cells making up the living tissue.

The heart

The heart is a muscular pump with two sides and four chambers, two on each side (Fig. 5.1). The upper chambers are called auricles and the lower chambers ventricles. The auricle and ventricle on each side are separated by very effective valves. The flow of blood (Fig. 5.2) through the heart is as follows:

1. Blood from the body enters the right auricle as the walls of this chamber expand.
2. Blood passes through the tricuspid valve to the right ventricle.
3. The right ventricle contracts and blood is pumped into the pulmonary artery leading to the lungs.

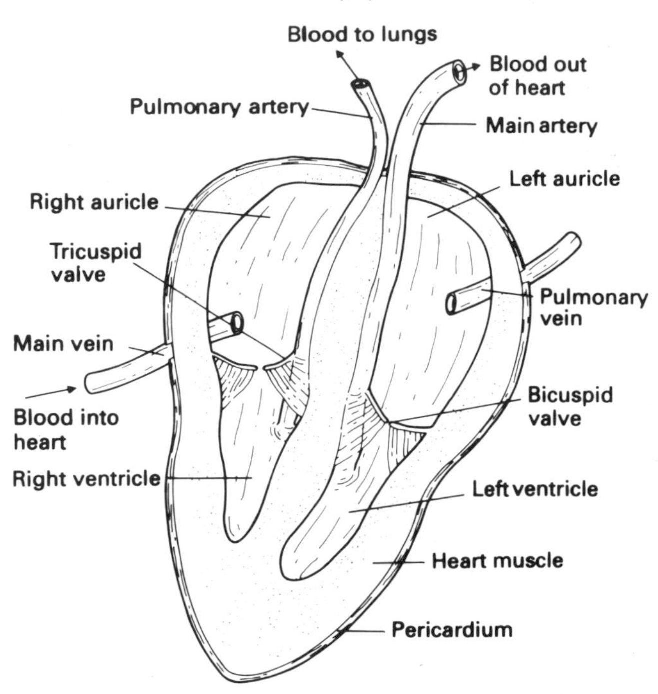

FIG. 5.1 Cross-section through a heart.

4. After the blood has released carbon dioxide and collected oxygen from the lungs the blood travels via the pulmonary vein to the left auricle.
5. It then passes through the bicuspid valve into the left ventricle which then pumps the blood through the aorta to all parts of the body.

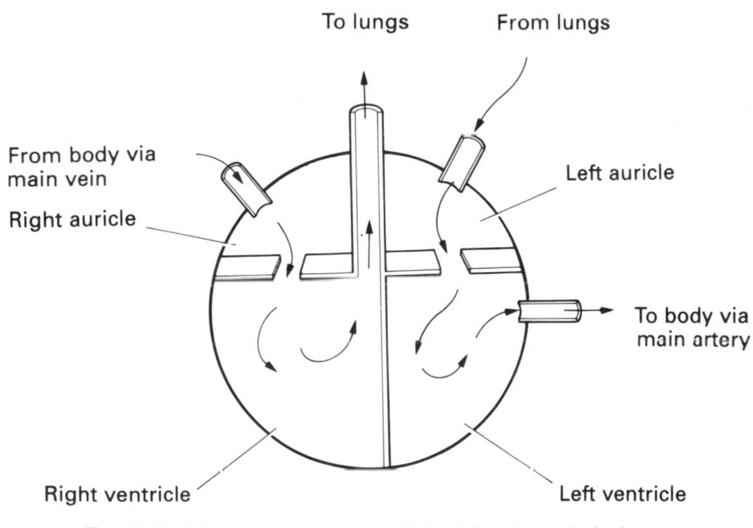

FIG. 5.2 Diagrammatic scheme of blood flow through the heart.

TABLE 5.1 *Temperature and pulse rates.*

Species	Normal temperature (°C)	Pulse rate per minute, at rest
Cattle	38.7	45–50
Sheep	40	70–90
Pig	39.7	70–80

Because of the force needed to pump blood around the body, the left ventricle has a much thicker, more muscular wall than other parts of the heart. The heart is contained in a sac called the pericardium.

Temperature and pulse rate

One of the functions of the circulatory system is to help maintain the body at an even temperature. Table 5.1 shows the temperature and pulse rates of cattle, sheep and pigs. The temperature of an animal is taken by inserting a clinical thermometer into the rectum. This should be held in position for the required time (usually one minute) and the reading taken. Illness is not the only cause of a rise in temperature. Exertion, hot weather and even handling the animal can cause a slight rise. The temperature of a young animal is often 0.5–1 Celsius degree higher than normal.

A pulse is the result of blood surging through an artery after a heartbeat. It is best detected where an artery is close to the skin surface and passes over a bone. The commonest places for taking the pulse are either under the jaw bone or on the inside of the fore limb. Unlike the temperature, the pulse rate of an animal is of little use to the stockperson.

The lymphatic system

The lymphatic system consists of a series of tubes throughout the body but, unlike the blood system, there is no pump. Lymph fluid comes from the blood plasma that passes out of the capillaries to help the functioning of individual cells. The lymphatic system eventually joins the main vein near the heart.

At various sites around the body there are lymph nodes which form lymphocytes. Any invading bacteria that are not cleared from the site of entry will find their way to the lymph nodes, which act as a further defence mechanism. The lymph nodes became large and inflamed when an animal is infected with disease, and this fact is used by the meat inspector at the slaughterhouse. The inspector will examine the lymph nodes of a carcass and if they are swollen he/she will know that the animal is or was recently ill and thus the carcass will demand closer inspection and could be condemned.

As has been mentioned in Chapter 3, the lymphatic system is also involved with the absorption of fats from the small intestine.

Chapter 6

The excretory system

Normal cellular activity results in the formation of waste products. In large quantities some of these products can be toxic and need to be excreted. The waste products that require excretion are:

Carbon dioxide.
Water.
Salts.
Nitrogen products (converted to urea in the liver).

These are transported in the blood system to the organs which make up the excretory system—the lungs, the skin and the main excretory organ, the kidneys.

The lungs

These have been dealt with in Chapter 4—The Respiratory System. Carbon dioxide and some water diffuse from the blood stream and are excreted by the lungs.

The skin

The skin is a protective outer layer to the body (Fig. 6.1). It has many nerve endings which transmit the sensations of heat, cold and touch to the brain. The outer layer, the epidermis, is a mass of cells with no blood supply or nerve endings. Its main function is protection and it varies in thickness according to the area that needs protection. Below the epidermis is the true skin, the dermis, with a blood supply and nerve endings. It contains hair follicles and sebaceous or grease glands which produce a protective film of oil. This oil, such as the oil lanolin in sheep's wool, protects the coat from water penetration.

The skin also contains many sweat glands. Each gland consists of a tube with one end opening as a pore on the skin surface, the other end being blind. The tube is surrounded by blood capillaries. Water and salts diffuse out from the capillaries and into the sweat tube. This travels to the skin surface where it evaporates to aid cooling of the body.

Sweat glands are present in the skin of all livestock except the goat. Cattle and pigs sweat only from the nose and snout, whilst a sheep has a thin supply of sweat glands all over its body. So much for the saying 'sweating like a pig'.

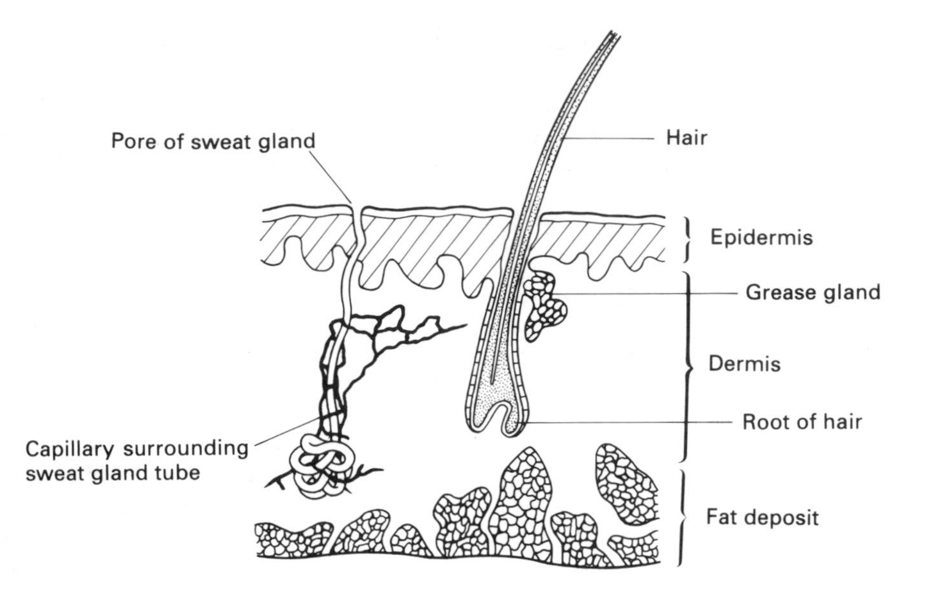

FIG. 6.1 Cross-section through skin.

Body heat control

Warm-blooded and cold-blooded animals react in opposite ways to changes in temperature. As the temperature falls, the cold-blooded animal becomes less and less active because it cools with the surroundings and its metabolic processes slow down as the temperature falls. The ultimate in this process is hibernation. As the temperature rises, the cold-blooded creature becomes more and more active, as with insects on a hot day. This speeding up of activity could go on until a temperature was reached at which life was impossible.

The warm-blooded animal keeps an internal temperature which is almost constant and in cold conditions must, above all else, prevent its internal temperature dropping. In conditions of high temperature the warm-blooded animal becomes less active in order to avoid generating more heat.

The source of heat in the animal body is the oxidation of food to provide energy. Heat is lost by radiation from the body, convection, i.e. in the air passing over the animal and by contact, particularly with the floor or ground when lying down. Heat loss is reduced by the presence of hair, feathers and subcutaneous fat. The control of body temperature is best followed by considering the two opposite alternatives, temperature falling and temperature rising.

Environmental temperature decreases

A fall in ambient temperature can be countered by increasing the metabolic rate. Animals that are tied will stamp and shiver; groups of loose animals may run around, but are more prone to seek protection behind a hedge or in the warmest part of a building and then will lie down close together. Pigs will fight for the central position and cold buildings can sometimes initiate the vice of tail-biting.

The involuntary physiological control of internal temperature lies in cutting down the blood supply to the surface by constriction of the arterioles to the skin. Under more severe conditions the blood supply to the extremities of ears and limbs is reduced.

Environmental temperature increases

The animal cannot avoid generating heat, though by resting this may be reduced to a minimum. The arterioles of the skin are now dilated. Blood flows freely and the skin is kept at a high temperature. Provided the environmental temperature is lower than that of the animal's blood, a condition which always applies in a temperate climate unless in a building where the temperature is abnormally high, heat will be lost by this means. In hot conditions this loss may be small and less than the internal production of heat. The sweat glands will then start secreting. Except in very humid, still, conditions this disposes of a lot of heat especially in those animals plentifully supplied with sweat glands, for the cooling effect of evaporation is enormous.

Most farm animals, however, are not efficient sweaters, but they have a technique which results in the evaporation of water. They breathe very rapidly, causing a lot of air to pass rapidly in and out of the mouth. Sheep and cattle pant in this way, but the dog is the most efficient because it has a relatively large, wide-opening mouth and a highly mobile tongue which is allowed to flop out. The mouth and tongue are kept very moist. This is a technique which man cannot imitate, for he always over-ventilates the lungs, removes carbon dioxide too rapidly and becomes dizzy.

The animals who pant to dispose of heat only allow a very little new air to enter the lungs each breath so that over-ventilation does not take place, but the dead space in respiration from the bronchi to the lips is continually receiving a fresh supply of air on to a surface kept wet from within.

The pig is not an efficient sweater, nor has it the power of effectively disposing of heat by panting. Its method of cooling is to wallow in water or mud. When it is in a very hot building and cannot follow this instinct it will urinate on the floor and wallow in its own urine, first on one side then the other, thus continuously exposing a wet side to the air.

If heat regulation fails and the animal's temperature rises, a very dangerous situation arises. Heat stroke may occur and death is likely to follow. In heat stroke the nervous regulation of heat breaks down. In hot weather animals tend to eat less, an important point, for digestion generates a lot of heat. Exercise and excitement also increase the heat and the problem of heat disposal.

The kidneys

The two kidneys are the main excretory organs of the body. In sheep and pigs they are the normal 'bean' shape and are held on either side of the spine in the lumbar region. In cattle the kidneys have a lobular structure (Fig. 6.2), with the left kidney

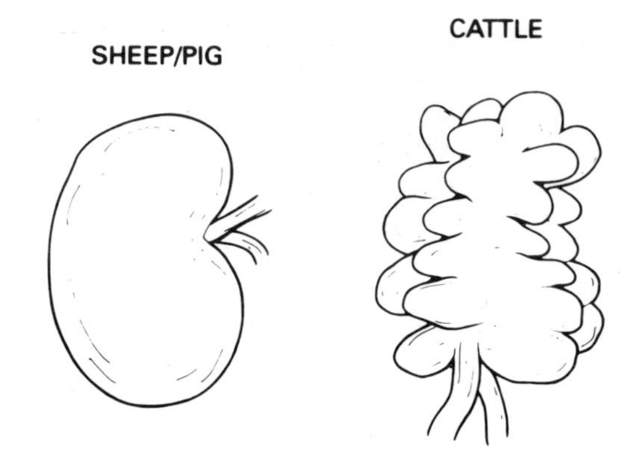

FIG. 6.2 External shape of a sheep/pig's kidney compared to that of cattle.

being able to 'float' in order to accommodate the changes in shape of the rumen. They are situated on either side of the spine in the lumbar region.

Kidney tissue consists of two distinct areas, the outer cortex and the inner medulla. Each kidney is supplied with blood from the renal artery, with the renal vein taking blood away (Fig. 6.3). There is also a tube called the ureter which takes the mixture of excretory products, the urine, to the bladder.

Each kidney has many urine-collecting tubules arranged radially. A cup-shaped structure called Bowman's capsule (Fig. 6.4) forms one end of the tubule, the other end opening into the ureter. The renal artery supplies each kidney with blood that contains water, salts and urea that need to be removed. Inside the kidney the renal artery divides into smaller branches, with each branch forming a small bunch of capillaries inside the Bowman's capsule. This bunch of capillaries is called the glomerulus. Waste products in the blood diffuse out of the capillary walls into the capsule, with the resulting urine passing into the ureter.

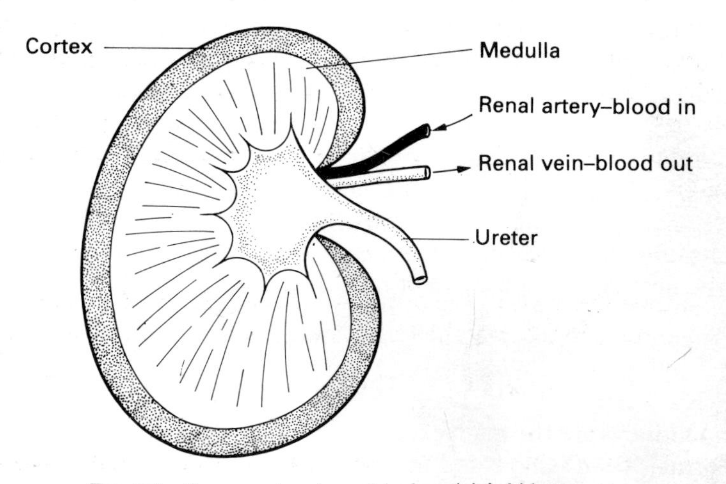

FIG. 6.3 Cross-section through a sheep/pig's kidney.

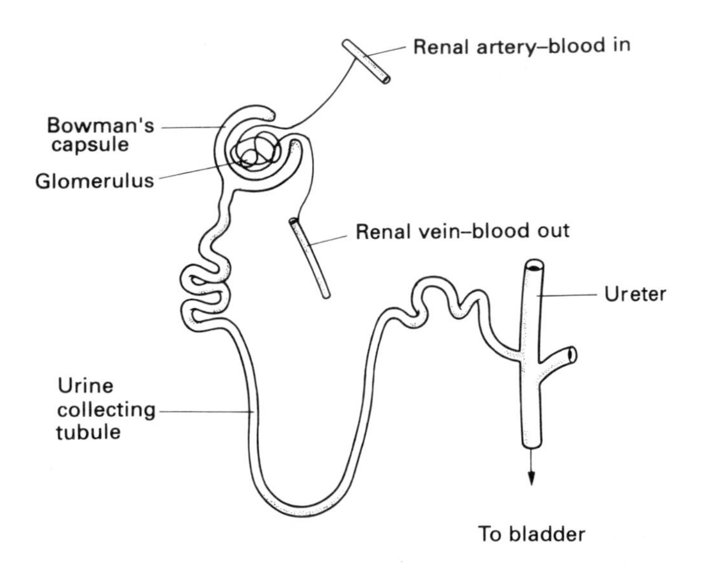

FIG. 6.4 The urine excretory system.

Although under the overall control of hormones, the kidneys are also responsible for maintaining the balance of water in the body. As urine passes down the tubule the required amount of water is absorbed back into the body.

The capillaries from each Bowman's capsule rejoin to form the renal vein which now carries blood that is cleaned of excess salts, water and urea.

The ureters and the bladder

Urine is continually in production in the kidneys and trickles down the ureter of each kidney to enter the bladder situated on the floor of the pelvis. The bladder is a collecting cistern composed of highly elastic muscular tissue. When empty it is quite small, but when distended it extends well forward from the pelvic girdle. The bladder is pear-shaped, with the narrow end pointing hindwards, ending in a sphincter muscle which leads into the urethra. Normally the sphincter muscle is closed, but as the bladder becomes distended nerve impulses lead to the releasing of the sphincter and at the same time contraction of the muscles of the bladder wall to assist in the elimination of the urine.

The urethra

No sex differences exist in the organs described so far. The urinary system is not part of the genital system, but the two systems share common terminal passages and openings.

The female urethra

The urethra in the female is short, with a wide passage, and opens into the vagina just above the bladder. The vagina is the common passage with the genital system which opens to the exterior through the vulva, an opening situated immediately below the anus.

The male urethra

In the male the urethra is a long narrow tube directed backwards from the bladder along the floor of the pelvis and then curled down and forward under the pelvis to continue in the substance of the penis.

Chapter 7

The endocrine and nervous systems

There is a very close link between the endocrine glands and the nervous system. Indeed one of the major glands in the body, the pituitary gland, is situated at the base of the brain, closely connected with the brain stem—the hypothalamus. Some responses of the nervous system are a result of glandular secretions, i.e. hormones, whilst some glands produce hormones as a response to nerves being stimulated.

Type of action

1. Glands—produce minute quantities of hormone which act reasonably quickly and their action can be inhibited quickly.
2. Nerves—when stimulated will produce an instant response over which the animal has complete control.

The endocrine organs

These are also called ductless glands. They produce secretions known as hormones that pass directly into the blood stream instead of being led by ducts to the areas where they are required. Once in the blood stream, they can travel to all areas of the body as necessary.

These glands are situated throughout the body (Fig. 7.1).

The glands and their main function are as follows, with a summary provided in Table 7.1.

Thyroid gland

The thyroid gland is situated in the neck close to the larynx and trachea. Thyroxin, the hormone produced by this organ, contains the element iodine which is essential for its production. Thyroxin is a general long-term metabolic stimulant, that is to say it produces no sudden burst of activity, but quite literally raises the rate of living. Thyroid deficiency results in sluggishness, obesity and hair loss. Thyroid excess

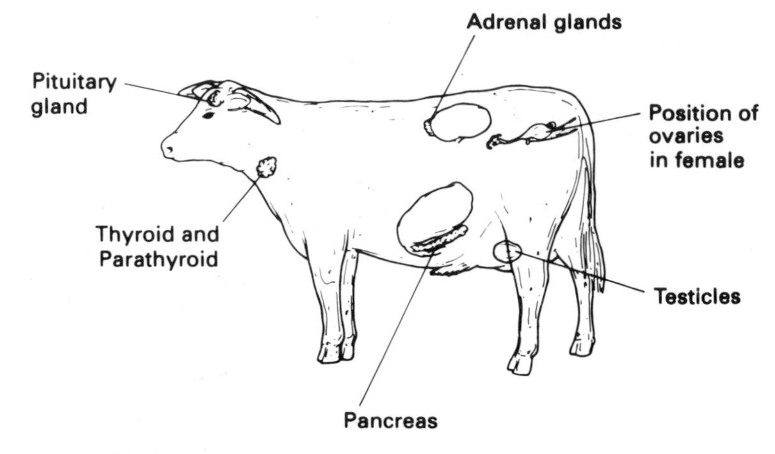

FIG. 7.1 Position of endocrine glands in body of cow/bull.

increases growth rate, activity and appetite and also leads to early senility. In thyroid deficiency, mental and sexual development are markedly impaired.

Hypothyroidism (inadequate thyroxin production) is liable to occur in areas where the soil has a low iodine content. Such areas occur widely throughout the world, and include substantial areas of North America, parts of the Swiss Alps and, in Britain, the county of Derbyshire. Hypothyroidism can easily be prevented by supplying small quantities of iodine, but this must be carefully controlled as iodine is a dangerous cumulative poison. Correct levels of iodine in the ration are most important for the correct functioning of thyroxin.

Parathyroid glands

These are small reddish-brown glands found either on or near the thyroid gland. There are usually two or four in number. They produce parathyroid hormone which helps to control the calcium content of the blood. As low levels of blood calcium are responsible for milk fever, this gland and its hormone can help to reduce the incidence of this problem. One of the preventive measures for milk fever is to inject the cow with vitamin D two to three days before calving, as vitamin D helps to increase the action of parathyroid hormone and thus elevates the levels of calcium in the blood.

Adrenal glands

These are small paired glands situated beside each kidney. They consist of two parts.

1. The cortex—the outer part, produces hormones known as corticosteroids. These have three main functions:
   i. The control of the metabolism of minerals.
   ii. The control of the metabolism of carbohydrate. The problem of acetonaemia (or ketosis) is a result of a build-up of toxic ketone bodies in the blood due to the

TABLE 7.1 *Hormones and their principal actions.*

Organ	Hormone	Action
Thyroid	Thyroxin	Metabolic stimulant
Parathyroid	Parathyroid hormone	Increases the calcium content of the blood. Decreases the phosphate content of the blood.
Adrenal glands		
1. Cortex	Corticosteroids	1. Control of mineral metabolism especially sodium. 2. Control of carbohydrate metabolism. 3. Reducing sensitivity and immunity reaction.
2. Medulla	Adrenalin	Conversion of glycogen into glucose and lactic acid. Rise in heart rate and blood-pressure. Relaxation of digestive system. Dilatation of the bronchi. Dilatation of pupils and tensing of eyeballs. Hair and feathers on end; sweating.
Pancreas	Insulin	Storage of blood glucose as glycogen.
Pituitary gland	1. Growth hormone	Stimulates growth.
	2. Hormones to the other endocrine glands	Stimulates the activity of the respective glands.
	3. FSH	In the female Enlargement of ovarian follicle. In the male Sperm production.
	4. LH	In the female Maternal behaviour Ovulation and formation of corpus luteum. In the male Development of testicles.
	5. Oxytocin	Milk 'let-down'.
Ovary		
1. Follicle	Oestrogen	Development of female conformation, organs and sexual behaviour of oestrus.
2. Corpus luteum	Progesterone	Development of uterus for pregnancy, udder for milk production. Depresses FSH and LH
Testicle	Testosterone	Development of male conformation, organs and behaviour.

breakdown of certain forms of carbohydrate. When the body is functioning normally, corticosteroids keep this problem under control.

iii. They reduce the sensitivity of the body and the body's reaction to outside organisms.

The latter point (iii) is widely practised in both veterinary and human medicine to help the body overcome diseases by preventing over-reaction to disease and so help reduce reaction to transplant surgery.

2. The medulla—The inner part of the gland which produces the hormone adrenalin. This is the so-called 'fright, flight and fight' hormone, in that it prepares the body for reaction to stress and to dangerous situations.

The pancreas

Much of the tissue of the pancreas is concerned with digestion and in this respect the gland has a duct, but there are sections of tissue known as the islets of Langerhans which do not discharge into the duct but act as an endocrine organ. They produce insulin, which has the effect of converting the glucose circulating in the blood into glycogen, an insoluble carbohydrate which is stored as a reserve in the liver and muscles. Thus insulin and adrenalin act in an antagonistic way which produces a balance. Insulin arranges the storage of potential energy, while adrenalin makes it available.

In the absence of insulin, sugar accumulates in the blood and begins to be excreted by the kidneys in the urine. Degeneration of the islets of Langerhans is no problem in farm livestock, but it is a common failure in man. It results in diabetes, a disease which was often fatal until the discovery over fifty years ago of the properties of insulin. Regular administration of insulin or a similar synthetic product can counter the loss indefinitely.

The pituitary gland

The pituitary is a small gland situated at the base of the brain. It produces a whole series of hormones, most of which are concerned with stimulating other endocrine organs. These hormones largely control the development of the reproductive organs, the reproductive cycle, successful pregnancy, the development of the udder and the 'let-down' of milk (*see* page 64).

Another hormone produced by the pituitary gland is growth hormone. It has little effect on the reproductive organs, but the rate of its secretion during the growth period is largely responsible for the ultimate size of the individual.

The reproductive hormones produced by the pituitary gland

The role of the reproductive hormones of the pituitary and how they help to control the oestrus cycle is dealt with in Chapter 8. The following is a list of the hormones produced and a brief description of their action.

Follicle stimulating hormone (FSH)

Before puberty, FSH is involved with the growth of the male and female reproductive organs. After puberty, FSH acts on the ovary to bring an ovarian follicle to maturity.

Luteinizing hormone (LH)

This causes the mature follicle on the ovary to burst, releasing an ovum (or egg). It then supplies the stimulus for the production of the corpus luteum (or yellow body) on the site of the ruptured follicle.

In the male, FSH is responsible for the development of sperm and LH stimulates the growth of the supporting tissue in the testicles where the sperm are produced.

Oxytocin

This hormone has a number of effects on the animal, but is primarily associated with milk 'let-down'. Any noise or action associated with suckling or milking, such as the noise of the milking machines or the action of the calf butting the cow's udder will result in milk being released form the milk-secreting cells into the udder.

Hormones produced by the reproductive organs

Female

Oestrogen

This hormone causes oestrus. It is found in the follicular fluid of the ovary. It is produced in small quantities in early life, even in fetal life, and results in female conformation.

Progesterone

Produced by the corpus luteum. If conception takes place, the corpus luteum, producing progesterone, persists and acts as a stimulus to enlargement of the uterus and also leads to an increase in milk-secreting tissue in the udder.

Progesterone has an antagonistic effect on the pituitary gland, tending to suppress the production of FSH and LH so that during pregnancy no follicles will ripen and mature on the ovary.

Prostaglandins

Produced by the reproductive organs of both females and males. Prostaglandins have many functions, but their most common use for farm animals is to induce oestrus, calving, lambing and especially farrowing.

Male

Testosterone

Produced by the testicles and is responsible for the development of male characteristics such as conformation and behaviour.

Prostaglandins

See section on female hormones.

The nervous system

The 'central nervous system' is the name given to the brain and spinal cord; the rest of the nerves are described as peripheral. The cells of the nervous system have processes which carry impulses or messages. There may be several short processes to receive messages, but only one process takes impulses from the cells, and this process may be extremely long, extending right down a limb, for instance. Nerve cells are not larger than other cells and these long processes are extremely fine individually, though large numbers of them are bound together to form the nerve which can be seen by the naked eye.

The central nervous system

All conscious action and thought is centred in the right and left hemispheres of the brain. The apparently simply actions of the animal body require a complex organization and many different tissues are involved. Many muscles have to be co-ordinated in moving a limb and many more in the relatively complex action of walking. By the time an animal can walk and trot easily, the individual actions are all controlled subconsciously, the higher part of the brain merely giving general instructions. Many actions are automatic, the act of breathing for instance, even though the individual is conscious of breathing and can, within limits, exercise control over it. Most activity is in response to a physical stimulus. An animal twitches its skin in response to the irritation caused by a fly. The action follows very quickly on the cause, but there is a slight delay while the incoming impulse travels to the brain and the consequent outgoing message is sent to the muscles or other organs involved.

The brain and spinal cord are protected by three serous membranes known as the meninges.

The peripheral nerves

Nerves leave the brain to serve the head and the special organs of sight, hearing and smell. Branches leave the spinal cord behind the arch of each vertebra in pairs, one each side. The main nerve trunk of the limbs, such as the radial in the fore limb and the sciatic in the hind limb, are derived from several branches of the spinal cord, junction occurring before they enter the limb. All nerves are composed of bundles of nerve processes in enormous numbers. The nerves in these bundles are of two types, afferent nerves bringing messages to the brain, sensations of touch, temperature, etc., and efferent nerves taking messages from the brain which results in action. Afferent nerves may serve a large area where detail is not important, i.e. the back, but in areas where detail is important, i.e. the human finger, each nerve serves only a small area. All parts of the body are served by both afferent and efferent nerves.

The autonomous nervous system

In addition to nerves serving centres in the brain, a whole system of nerves governs the involuntary activities of the body. These autonomous nerves are linked to each other and have connections with the spinal cord. They control the activities of glands, the involuntary muscles of the digestive and other systems, the heart muscle and the pupil of the eye.

The autonomous nerves are in two groups working against each other to meet the varying needs of the body. One group, the sympathetic nervous system, produces effects similar to those produced by the hormone adrenalin. It reduces the activity of the digestive system, stimulates the heart muscle and tends to concentrate the blood in the areas of action. The sympathetic nerves are therefore in the ascendancy in emergency. The other group, the parasympathetic nerves, do not act together in the same way, because they stimulate internal activities which have no essential connection. These nerves stimulate muscular and glandular activities in the digestive system, urinary and sexual activity and have a depressing effect on the heart muscle.

Chapter 8

The reproductive system

The basis of many stock farming systems is to ensure that the animals concerned have a regular reproductive cycle. Stockpeople spend large amounts of time watching for animals coming on heat, assisting at births and getting the animals back in-calf, in-pig or in-lamb.

To help with this vital part of the stockperson's job a thorough understanding of the male and female reproductive systems is essential.

The male reproductive organs

The male reproductive organs consist of two testicles (testes), a penis and accessory structures (Fig. 8.1). Sperm for reproduction is produced in the two testicles which are situated in the scrotal sac. Each testicle is divided into compartments, or lobes, containing specialist sperm-producing cells. Once produced, the sperm are stored in a long coiled tube known as the epididymis. Nerves and blood vessels serve each

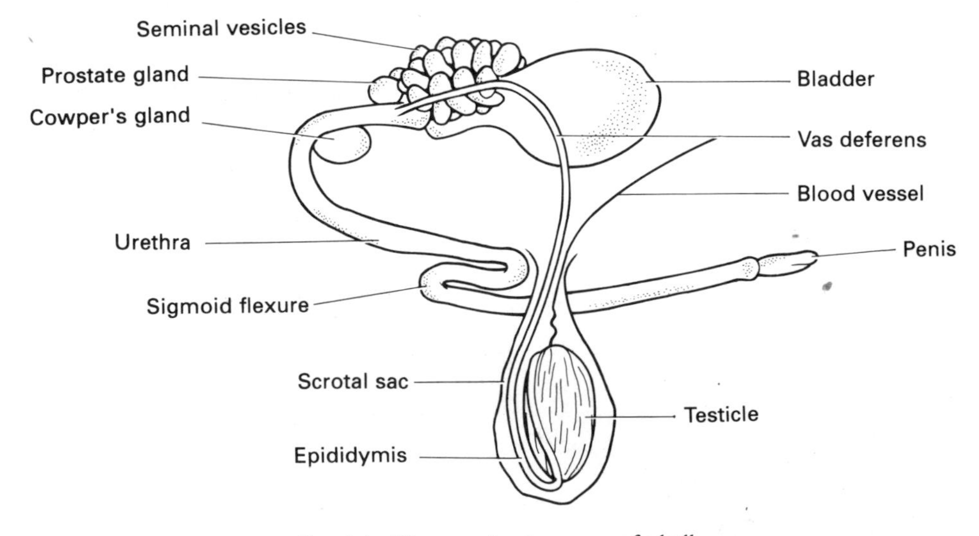

FIG. 8.1 The reproductive organs of a bull.

testicle, as does the vas deferens, the tube along which the sperm will travel from the epididymis to the urethra. It is the vas deferens that is cut when a veterinary surgeon performs a vasectomy to produce, for example, the teaser ram. The animal retains its male characteristics but is infertile.

At the junction of the urethra and the vas deferens there are various glands, notably the seminal vesicles, the prostate gland and Cowper's gland. These produce semen which mixes with the sperm. The function of the semen is to provide food for the sperm and to provide an alkaline medium for the sperm to travel in. The penis is normally held in a protective fold of skin known as the prepuce.

In order to keep the testicles cooler than the rest of the body (see below—problems affecting fertility), the scrotal sac is suspended under the abdomen. In the boar the scrotal sac is held close to the body just below the anus.

Problems affecting fertility

There are two main aspects of this—firstly the animal may produce infertile sperm and secondly he may have no sex drive.

Temperature

The testicles need to be kept 3–4 °C cooler than the body temperature for efficient sperm production. Thus a spell of hot weather can reduce the fertility of some farm animals, especially boars. Rams with a heavy growth of wool on the scrotal sac may suffer from a reduction in fertility.

Nutrition

Underfeeding or overfeeding can lead to fertility problems. If underfed, the animal's body has a higher priority for its food than sperm production, whilst overfeeding can lead to a fat, lazy animal with little sex drive.

Incorrect feeding is probably the most common cause of infertility in the male.

Work load

An overworked animal will have fewer sperm per ejaculation, resulting in reduced fertility. It will also suffer a reduced sex drive. The recommended work loads for bulls, boars and rams are shown in Table 8.1.

TABLE 8.1 *Recommended work loads for bull, boar, and ram.*

	Services per week	
Species	Young breeding male	Mature breeding male
Bull	1–3	4–5
Boar	2–3	4–5
Ram*		12–15

* These figures relate to rams standing at artificial insemination centres. Frequency of services for rams varies considerably with time of year (breeding season) and ewe management system (whether ewes are synchronized or not).

Disease

Infections of the reproductive tract can result in inflammation of any parts of the system, resulting in temporary or permanent infertility. These infections include brucellosis and trichomoniasis, which can be transmitted from male to female or female to male during natural service (*see* Chapter 11).

Housing

Animals kept in unsuitable environments can become bored, leading to loss of sex drive. This sometimes happens to bulls kept in poorly designed pens.

Physical abnormalities affecting fertility

Hernia

In the fetus the testicles originate near the kidneys, but slowly travel down to pass through an opening in the abdominal muscle wall at about the time of birth. If part of the abdominal contents, i.e. the intestines, also pass through this opening into the scrotal sac at the same time, the result is a hernia. This is a fairly common condition in pigs. A hernia may not necessarily adversely affect the fertility of the animal.

Rigg

A term used to describe an animal that is born with only one or even no testicles. It may be that they have failed to descend from the abdomen and may do so at a later date. This can cause problems if the animal is to be castrated as the testicle may descend after the legal age limit for the operation has passed.

Castration

Castrating animals is not as widely practised now as in past years. The faster growth rates and leaner carcasses achieved by entire males outweigh many of the advantages of castration. However, the operation is still performed on many farms on cattle, pigs and sheep. There are several methods of castration, each having its advantages and disadvantages, each having a set of legal requirements and each requiring a different practical technique. The technique of castration is best taught by a qualified person in a practical situation and so will not be dealt with in this book.

Bulls

1. The Burdizzo method, or bloodless castration. This method uses a large pair of 'pliers' called Burdizzos (named after its inventor). Two blunt crushing surfaces cut the blood vessel and sperm chord inside the scrotal sac. The scrotal sac itself is not cut. The result is that eventually the two testes shrivel and die. This method seems to cause great pain to the animal after the operation, with swelling of the scrotal sac that does

not subside for two to three weeks. There is little risk of infection however, as no incision has been made. A farmer may use this method of castration until the bull is two months old. After this a veterinary surgeon is required to carry it out. Animals need checking three to four weeks after the operation, as this method is not always one hundred per cent successful.

2. The rubber ring. A tight rubber ring is placed above the testicles which grips the neck of the scrotal sac. The effect of this is to cut off the blood supply and in time the testicles and scrotal sac shrivel up and fall away. This is a very convenient method, but is not allowed to be used after the animal is seven days old. There is a slight risk of infection, as a small open wound is left when the scrotal sac falls away.

3. Surgical castration. This methods involves an incision into the scrotal sac, cutting the sperm chord (vas deferens), pulling away the blood vessel and then removing each testicle in turn. It is a very 'sure' method, as animals are not usually missed or left half done. The pain involved for this method seems to be less after the operation than the two previous ones, but there is always the risk of infection from the open wound. The surgical method must not be used by a farmer after the animal is two months old, otherwise a veterinary surgeon is required.

Rams

The three methods of castration described for bulls also apply to rams. However, there is a slight difference in the legal aspects. A farmer cannot castrate a ram using the Burdizzo method or the surgical method after the animal has reached three months of age. The seven-day rule that applies to the rubber ring method remains the same.

Boars

As boars' testicles are held close to the body, the rubber ring method and the Burdizzo method are not practical. The surgical method of castration is the usual choice. Again a farmer is not allowed to castrate a pig that is over two months of age. The operations described above are allowed to be carried out by a farmer (or stockperson) within the time period specified without use of anaesthetic. As a general rule, operations outside the time period need to be performed by a veterinary surgeon using an anaesthetic. Refer to Table 14.1 in Chapter 14 for a summary of requirements for on-farm operations performed on various classes of livestock.

Artificial insemination (AI)

AI is one of the success stories of modern agriculture. It was first used in Italy as long ago as 1780; however, it was Russian scientists who developed the technique for farm animals, mainly cattle and sheep, over a thirty-year period from 1900. In the United Kingdom the Milk Marketing Board took on responsibility for creating a cattle AI service in 1944 and by 1951 an AI network covered England and Wales. The development in 1949 of freezing bull semen using solid carbon dioxide (−79°C) revolutionized AI techniques. However, by 1963 liquid nitrogen (−196°C) was being used to achieve lower temperatures and therefore longer semen storage times became possible.

AI offers some distinct advantages over natural service which are worthy of note:

Enables widespread use of top quality sires and therefore their genetic potential, so improving the performance of the national herd.
Allows the testing of progeny of many farms and so under different management systems in different environments.
Significantly reduces the spread of disease, especially sexually transmitted diseases.
Large groups of females can be served at the same time, thus enabling synchronization of the oestrus cycle.
Avoids the need to keep potentially dangerous bulls on the farm.
Can help preserve rare breeds.

The widespread use of a top quality sire is achieved by diluting the semen. Table 8.2 shows the number of inseminations (straws or doses) that can be provided, by dilution, from each ejaculate from cattle, pigs and sheep.

Techniques of insemination

Cattle

Deep-frozen bull semen is stored in straws which are used for insemination. The straws are quite thin, enabling them to be used for heifers and cows. Each straw, containing about 0.5 ml of semen, is loaded into an insemination 'gun'. The inseminator inserts his hand into the cow's rectum in order to manipulate the cervix (in the reproductive tract) and so guide the straw through the cervix. The semen is deposited in the body of the uterus, further up than if natural service is used.

Nowadays many farmers do their own inseminations (*see* Fig. 8.2), the so-called 'DIY AI'. During the intensive training for the technique, great emphasis is placed on the positioning of the catheter, as a heavy-handed insemination can result in the cervix being damaged.

Pigs

AI in pigs involves the use of a rubber or plastic catheter in order to insert the semen. Unlike the straws used for cattle, the catheter closely resembles the shape of the boar's penis. It has a spiral (corkscrew) tip which is inserted through the vagina

TABLE 8.2 *Number of inseminations per ejaculate.*

Species	Number of inseminations per ejaculate	Number of sperm per artificial insemination
Bull	400	5–15 million
Boar	15–30	2000 million
Ram	40–60	50 million

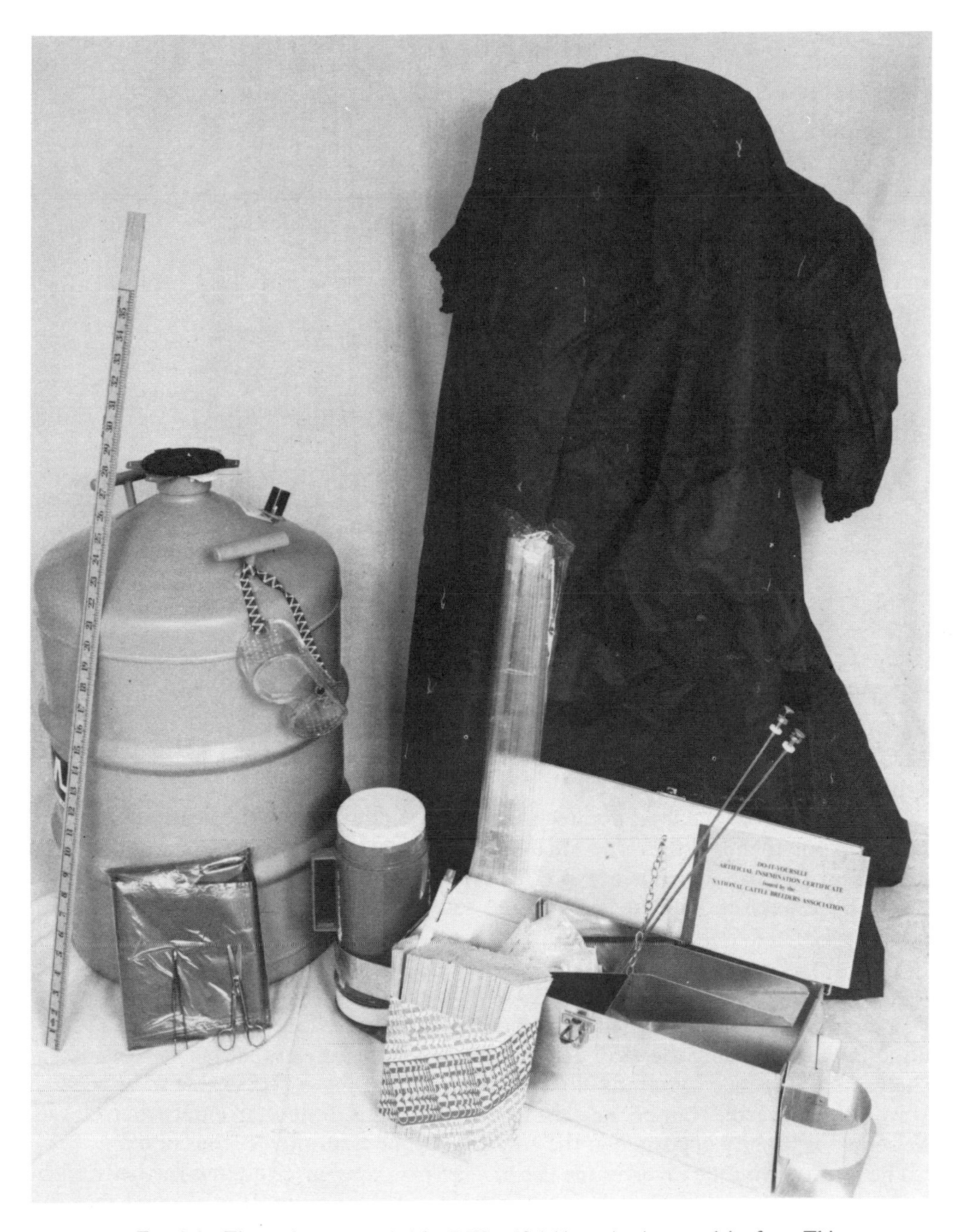

FIG. 8.2 The equipment needed for DIY artificial insemination on a dairy farm. This includes the storage tank for deep-frozen bull semen and the catheter that is inserted into the cow.

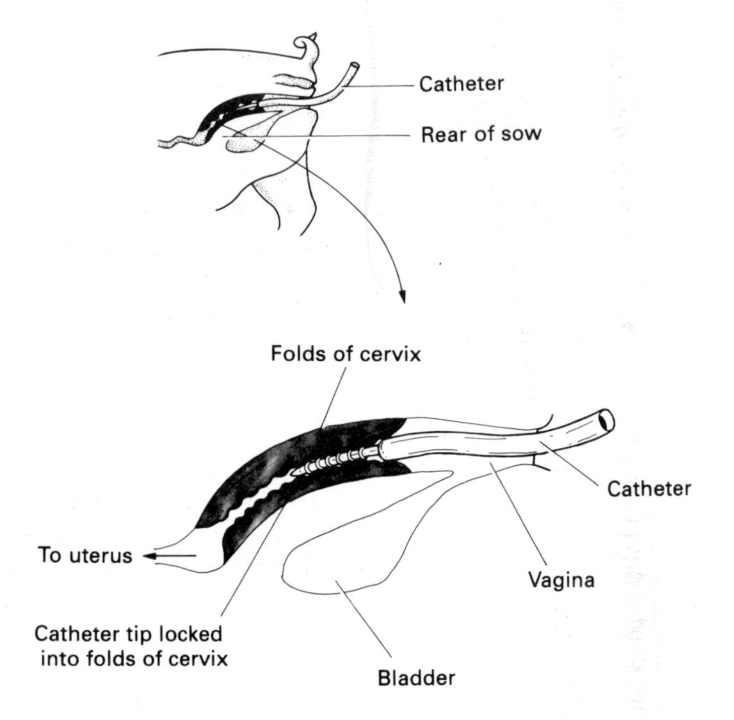

FIG. 8.3 Artificial insemination technique for sows.

and rotated in an anti-clockwise direction until it locks in the muscular folds of the cervix (Fig. 8.3). The semen is then deposited in roughly the same place as with natural service.

Commercial boar semen is not deep-frozen; it is delivered to the farm, diluted and cooled, in polythene bottles. The 'shelf life' of a bottle of boar semen is about five days. Deep-freezing boar semen is not as effective as with bull semen because of the wide variation between individual boars of the semen's sensitivity to preservation treatments.

Sheep

Frozen ram semen is available from commercial breeding companies. The technique involves holding the ewe, usually in a special crate, so that her head is down. The vagina is opened by means of a speculum so that the cervix is able to be seen. A straw is then inserted, but because the cervix of a ewe is difficult to enter when closed, the semen is usually deposited at the entrance to, or just outside, the cervix.

The latest technique involves the use of a laparoscope, inserted into the belly of the ewe, to place the semen directly into the horns of the uterus. It is claimed that this technique gives superior results to conventional AI.

The above are brief descriptions of the AI techniques for different species. The only way to learn the practicalities of the techniques is to go on an appropriate training course.

The female reproductive organs

These lie in the pelvic cavity and consist of two ovaries, two fallopian tubes, the uterus, the cervix, the vagina and the vulva (Fig. 8.4).

Ovaries

There are two ovaries which are oval in shape and vary in size (1.5–5 cm), depending on the species of animal and the stage of reproductive life of the animal. Inside each ovary are a large number of Graafian follicles, each containing an ovum (or egg). During the oestrus cycle a follicle or follicles will come to the surface of the ovary and form a blister-like structure. This will then burst, releasing an ovum (or ova if a number of follicles develop and burst).

Fallopian tube (oviduct)

These are two very narrow tubes which lead, one from each ovary, to the uterus. Once an ovum has been released by the ovary, finger-like strands (fibria) on the tips of the fallopian tube pick up the ovum, which is passed down the tube by means of cilia (hair-like structures) which set up a wave-like motion. If sperm are present, it is here that the ovum will be fertilized. The fertilized (or unfertilized) embryo (or ovum) then moves down to the uterus.

Uterus

The uterus is a hollow muscular organ consisting of a body and two horns. The size varies greatly, as it is here that the young develop. The length and shape of the horns

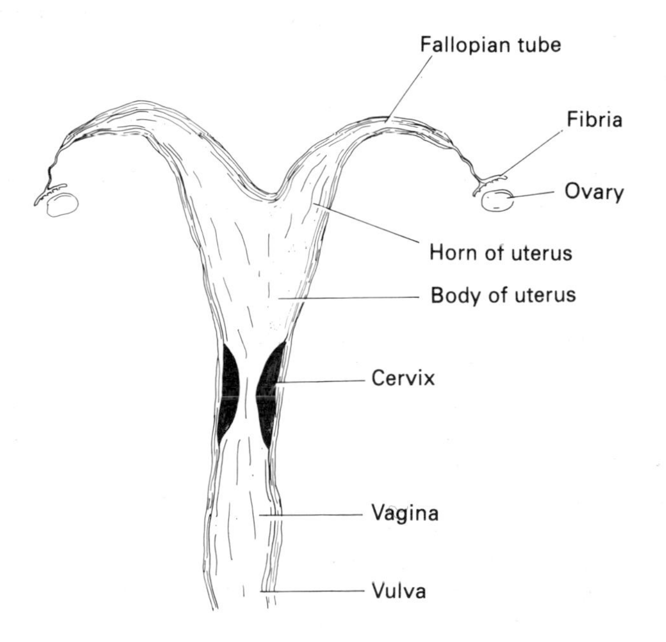

FIG. 8.4 The female reproductive organs.

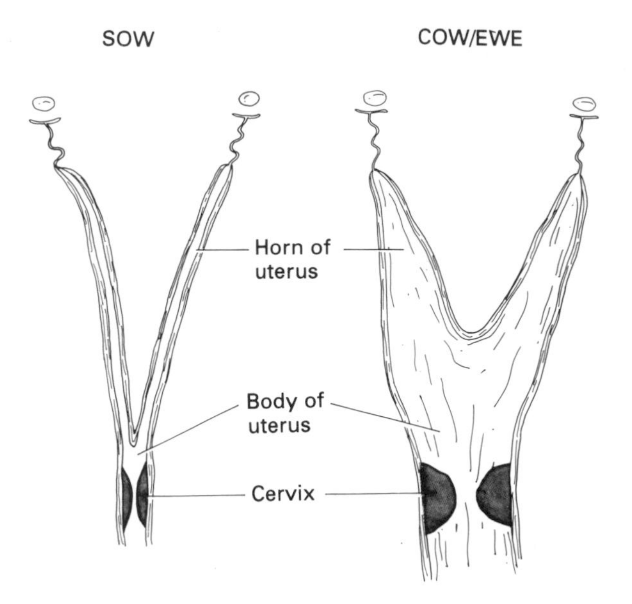

FIG. 8.5 Uterine horns of pigs compared to cattle and sheep.

differ according to species. Pigs have longer horns and a smaller uterine body than cattle or sheep (Fig. 8.5).

An embryo attaches itself to the wall of the uterus, the mode of attachment varying between the species. In single pregnancies the fetus lies in one horn and the body of the uterus. Twins may lie in one horn or one twin in each horn. In multiple pregnancies such as in the sow, both horns are occupied by embryos.

The end of the uterus is closed by a muscular area known as the cervix.

Cervix, vagina and vulva

The cervix, also known as the neck of the womb, is a thick muscular structure that forms a barrier between the external atmosphere and the uterus. A mucus plug is usually present during pregnancy and during part of the oestrus cycle, acting as a barrier to infection. The vagina extends from the cervix to the outside opening of the vulva. The walls are tough and elastic, containing mucus-producing cells. The urethra or urinary tract opens into the floor of the vagina. The vulva is the external opening to the reproductive system.

Hormonal control of the reproductive cycle

From puberty onwards the ovum (or ova) is shed at regular intervals according to the oestrus cycle of the species. For cattle and pigs it is twenty-one days, for sheep it is seventeen days. This cycle is interrupted only by pregnancy or occasionally by some malfunction of the hormones that control the cycle.

The description that follows is the sequence of events that control the oestrus cycle of cattle, but it is applicable to most common farm livestock:

1. Follicle stimulating hormone (FSH) produced by the pituitary gland stimulates follicles to develop on the ovary. One follicle, containing an ovum, will mature and form a blister on the ovary.
2. As this follicle develops, it produces oestrogen which has three main functions:
   i. To stimulate growth and development of the uterus ready to receive a fertilized ovum.
   ii. Rising levels of oestrogen begins to inhibit the production of FSH and stimulates the production of luteinizing hormone (LH) from the pituitary gland.
   iii. Oestrogen is responsible for the animal displaying the signs of heat that a stockperson will recognize, e.g. enlargement of vulva and the mounting behaviour of cows.
3. Increasing levels of LH cause the follicle to rupture releasing the ovum, i.e. ovulation.
4. Just after ovulation a gland begins to form exactly on the site of the ruptured follicle, growing until it protrudes from the surface of the ovary—this is known as the corpus luteum or, because of its colour, the 'yellow body'. This gland can be felt by the veterinary surgeon some five or six days after ovulation and, by its size, will give an indication of the stage of the oestrus cycle the animal has reached.
5. The corpus luteum produces progesterone which:
   i. Prepares the uterus for implantation of the developing ovum.
   ii. Suppresses the production of FSH and LH.

If the cow has been served and the egg is fertilized, the corpus luteum will remain throughout pregnancy, producing progesterone. If, on the other hand, she does not conceive, the corpus luteum begins to reduce in size and so the level of progesterone falls. The reducing level of progesterone signals the release of FSH and so another oestrus cycle will begin.

Both LH and FSH are manufactured by the pituitary gland, but their release is controlled by another hormone, gonadotrophin releasing hormone (Gn-RH), produced by the hypothalamus of the brain.

The reduction (or regression) of the corpus luteum is controlled by the hormone prostaglandin produced by the uterus.

Artificial control of the oestrus cycle

Artificial control of the oestrus cycle is achieved by the use of injections, coils and sponges that contain some of the hormones described above. It is done so that a veterinary surgeon can treat problems in individual animals or so a stockperson can synchronize ovulation in a group of animals.

1. Oestrogen is used to stimulate the ovaries to function when an animal fails to cycle after birth of the young. The problem with using oestrogen is that the animal may show signs of heat without an ova being released.

2. Gonadotrophin releasing hormone (Gn-RH) is used to stimulate the release of FSH and LH.
3. Luteinizing hormone is used on the day of service to ensure that ovulation occurs.
4. Follicle stimulating hormone (and pregnant mare serum gonadotrophin) is used in the technique of embryo transfer. Large doses of FSH stimulate several follicles to develop and so several ova are released (*see* page 53).
5. Prostaglandin. When administered as an intramuscular injection, it causes the destruction of the corpus luteum, so progesterone levels fall and Gn-RH is released to stimulate the production of FSH and LH, so bringing about the oestrus cycle. Prostaglandin will act on the corpus luteum whether the animal is pregnant or not. If the animal is pregnant, she will abort, so if this technique is to be used for cows the veterinary surgeon will often carry out a rectal examination to see if she is pregnant before injecting the drug.
6. Progesterone is usually administered in the form of a progesterone-impregnated coil or sponge. In sheep the vaginal sponge is used, in cattle the progesterone releasing intravaginal device (PRID). These are inserted into the vagina of the animal, progesterone is released from the sponge or PRID and absorbed into the blood stream. The progesterone blocks the activity of Gn-RH so no FSH or LH is released; however, the corpus luteum regresses quite naturally. When the device is removed, Gn-RH is activated, FSH and LH are released and the animal comes on heat two or three days later. These 'sponges' are used to synchronize the start of oestrus in sheep, cows and heifers, leading to more efficient timing of service and eliminating the need for heat detection.

   The low levels of progesterone in the blood and milk of farm animals can be used to detect pregnancy and heat. Various companies now produce kits that detect progesterone in cows' milk and in sows' blood. This works on the basis of the corpus luteum producing progesterone during the oestrus cycle for non-pregnant animals and producing progesterone throughout pregnancy for pregnant animals. (*See* Fig. 8.6 below and pregnancy detection on page 56.)

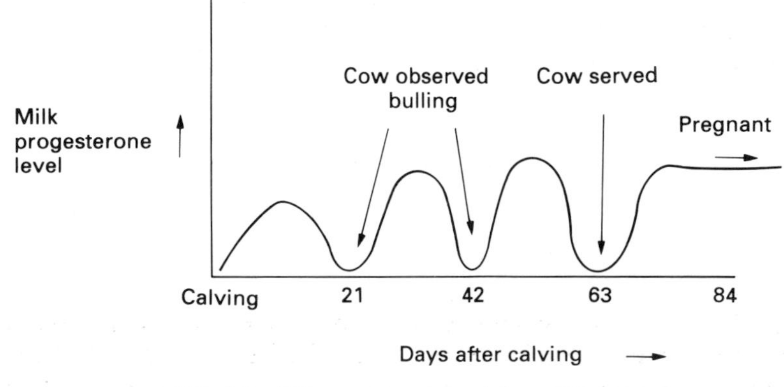

FIG. 8.6 Milk progesterone levels in a normal cow.

Embryo transfer

The technique of transferring a live embryo from a donor animal to a recipient animal is not new. It was first achieved by Heape in 1890 using rabbits. Nowadays cattle, sheep, pigs, goats and horses all have embryo transfer services available, with cattle having the major share of the commercial market in terms of number of transfers performed. An embryo transfer service is offered by the Milk Marketing Board of England and Wales and by a number of commercial companies in the United Kingdom.

In simple terms, the technique for cattle involves inducing the ovaries of a cow of high genetic potential to produce more than one ovum. This is achieved by injecting the cow with pregnant mare serum gonadotrophin (PMSG). The donor cow is inseminated, usually using a well-proven bull, and six to eight days after ovulation the embryos (i.e. fertilized ova), numbering on average four to six, are recovered from the uterus (see Fig. 8.7) using surgical or non-surgical procedures. The trend is away from using surgical procedures.

The embryos are checked for viability under a microscope and drawn either into a catheter for non-surgical transfer or into a pipette for surgical transfer. They are inserted into the uterine horns of recipient cows after ensuring that donor and recipient are at exactly the same stage of the oestrus cycle.

Normally only one to two per cent of natural cattle pregnancies result in twins, so embryo transfer can be used to ensure twin calves are produced. This can be done by adding an embryo to the uterus of a cow already carrying one embryo or by introducing two embryos into the recipient cow.

One problem with twins is that if they are of opposite sexes they may not develop full sexual maturity due to hormonal interaction whilst in the uterus, i.e. the condition known as Freemartinism. If the sex of the embryo can be determined before transfer into the recipient, this problem can be overcome. The technology does exist to determine the sex of embryos, but it is not commercially available in the United Kingdom. No doubt in the very near future sex determination will be commonplace.

To sum up, the advantages of embryo transfer are:

More calves are produced from the elite cows and bulls, so increasing the value of the herd.
It can be used to speed up herd breeding programmes dramatically.
It can ensure rapid multiplication of rare breeds.
It can increase the number of twins born if this is desirable.

Signs of oestrus

The cow

In a herd of cows the most common sign is to see one cow riding another, but the stockperson must check carefully to see which of the two cows involved is on heat. 'Spotting' cows on heat is one of the most important jobs on a dairy unit and every day

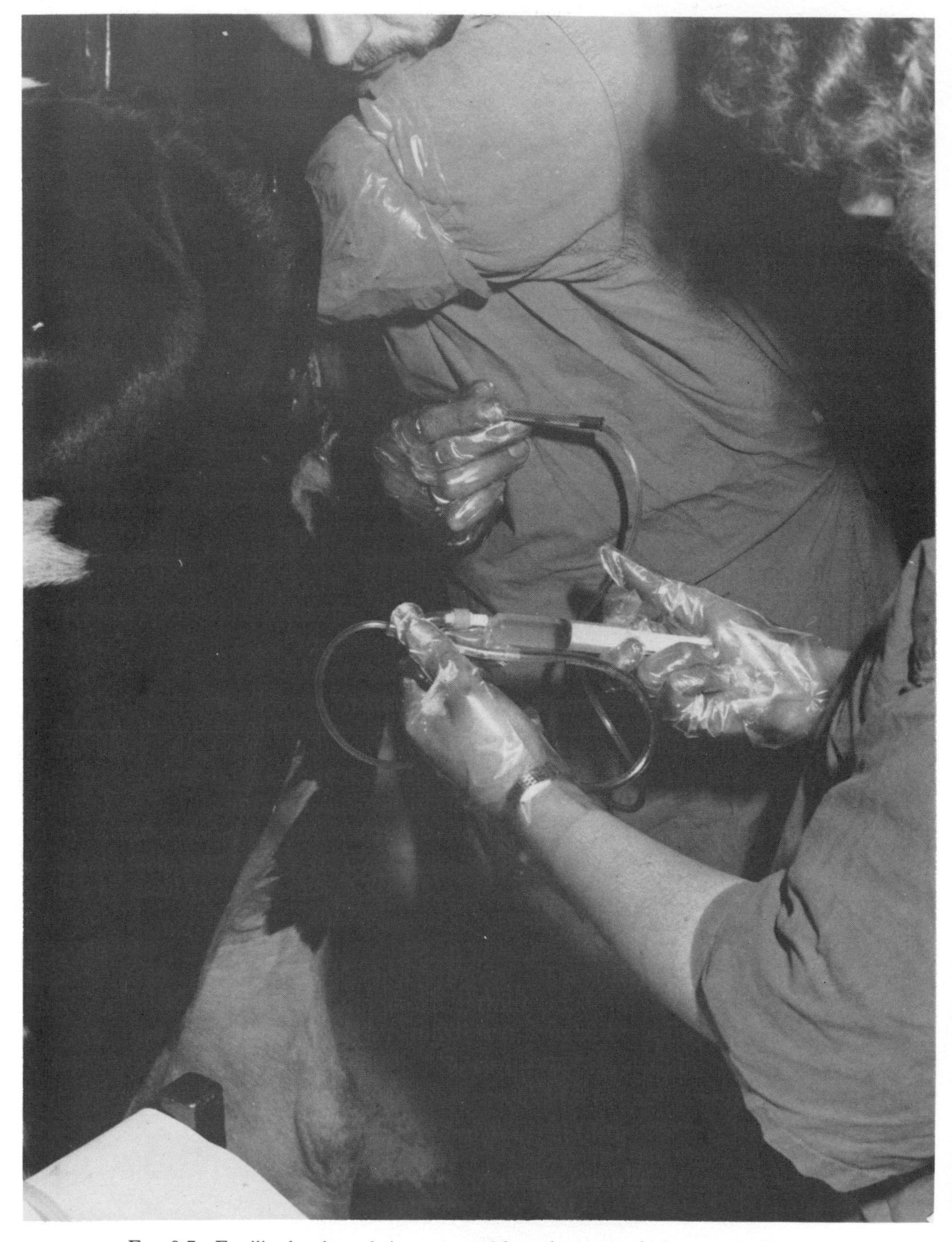

FIG. 8.7 Fertilized embryos being recovered from the uterus of a donor cow using the non-surgical technique.

time must be set aside for the stockperson to simply stand and watch the cows. Other signs include sniffing, chinning and a swollen vulva with a stringy mucus discharge.

The sow

A sow on heat is very docile. She will pace up and down outside a boar pen if allowed to do so and her vulva will be red and swollen, especially in gilts. The usual test is for the stockperson to press down on her back and shoulders. If she stands still for this back pressure, with her ears pricked back, then she will usually be on heat.

The ewe

It is very difficult to spot a ewe on heat unless a ram is present. He is attracted to her by smell and if on heat she will usually accept the ram without resistance.

Pregnancy

Once an ovum has been fertilized, the embryo travels to the uterus and implants itself in the uterus wall. The first outward sign of pregnancy is a break in the oestrus cycle. The abdominal enlargement due to the presence of the young in the uterus is remarkably difficult to judge, particularly in the cow, so should not be regarded as a reliable method of determining pregnancy. Provided the nutrition is correct and the animal is not put under any form of stress, pregnancy will normally progress satisfactorily. The gestation periods for farm animals are given in Table 8.3.

Any premature birth must be considered abnormal, although some degree of common sense must be applied to this judgement. Cows, for example, will often calve four or five days before or after the due date with no ill effects whatsoever. However, abortion in any animal is a serious matter, especially in cattle, as it could be caused by an infectious disease. It should be noted that, in cases of abortion in cattle, the stockperson must (by law) notify the Divisional Veterinary Officer for the area so that samples can be taken and examined for brucellosis (*see* page 80). This is normally done via the farm's own veterinary surgeon.

Stockpeople, especially pregnant women, should not handle aborted material from any animal, as there is a slight risk of picking up an infectious disease.

TABLE 8.3 *Gestation periods for farm animals in days and months.*

	Gestation length	
Species	Days	Months
Cow	282	9.2
Sow	115	3.7
Ewe	148	4.8

TABLE 8.4 *Common methods of pregnancy testing for farm animals.*

Method	cow	sow	ewe
Milk test, progesterone	√ (19)		
Blood test, progesterone		√ (19)	
Ultrasound, audio	√ (42)	√ (25)	√ (30)
Ultrasound, visual	√ (70)		√ (70)
Manipulation of uterus	√ (42)		

Figures in brackets are the earliest number of days post-service that pregnancy can be detected using that method.

Pregnancy detection

Pregnancy detection is a valuable aid to stockmanship, but it is no substitute for regular checking of stock for signs of returning to heat.

Table 8.4 shows the various pregnancy detection methods available, the species that each is applicable to and days after service when pregnancy can first be detected using that particular method.

Below is a brief description of how the 'on-farm' methods of pregnancy detection work:

Milk/blood progesterone

Test kits are available that consist of plastic tubes coated with chemicals that attract progesterone (see Fig. 8.8). A milk or blood sample is taken and put into the tubes. The tubes are then treated with various chemicals that will detect and colour any progesterone present. A colour comparison chart is provided and so long as the milk or

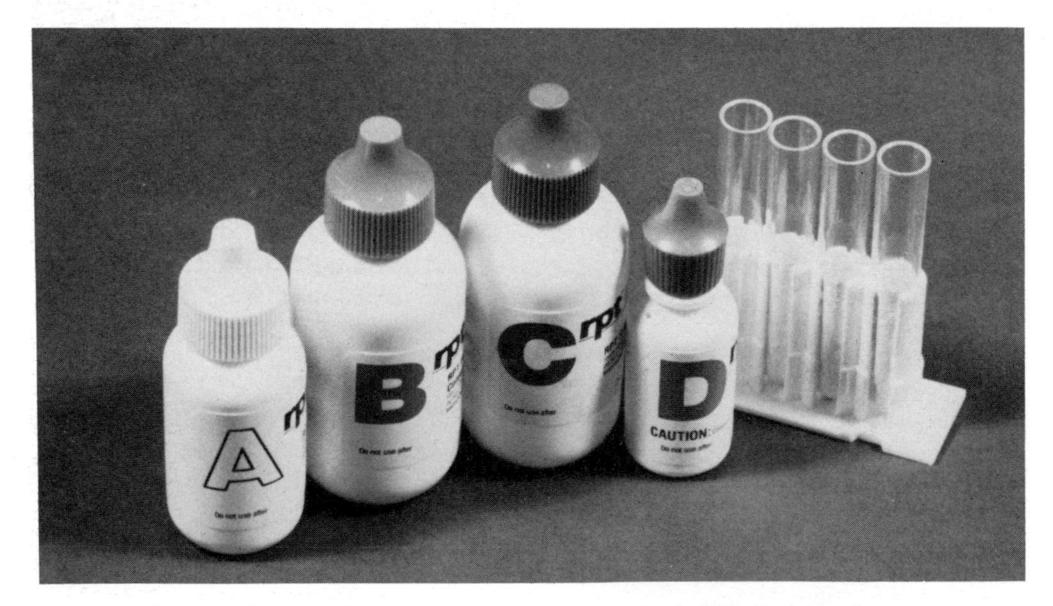

FIG. 8.8 The Rapid Progesterone Test kit available from the Milk Marketing Board.

blood sample has been taken at the correct time after service, the colour will indicate a high progesterone level (pregnant) or a low progesterone level (due to come on heat).

Manipulation of uterus

This is the most common method of pregnancy detection in cattle. The veterinary surgeon will feel and compare the size of the two uterine horns through the wall of the rectum. This method is very accurate and can detect pregnancy from about seven weeks after service.

Ultrasound scan, audio and visual

A machine to generate ultrasound (Fig. 8.9) is placed on the flank of the animal (sow/ewe) or inside the rectum (cow). The audio ultrasound scan can pick up the sound of the blood flow in the uterine artery or the fluid surrounding the fetus. The operator must have experience, as there is often a lot of background noise which has to be ignored as it can give a false reading to the untrained operator.

The visual ultrasound scan is a sophisticated piece of equipment usually used by specialist operators. The ultrasound scan builds a 'picture' of the fetus on a screen. Twins (in sheep and cattle) can be detected using this method.

Parturition

Parturition is defined as the act of birth, a general term covering calving in cows, lambing in ewes and farrowing in sows.

FIG. 8.9 Using an audio ultrasound machine to test for pregnancy in a sow.

As the female approaches the end of her pregnancy, the udder enlarges and milk will be secreted. The vulva becomes enlarged and up to forty-eight hours before parturition the ligaments at the rear of the pelvis will slacken. An animal in a field will tend to go off on her own and may stop feeding. A sow, if bedding is provided, will prepare a 'nest'.

Stages of labour

For theoretical purposes the process of giving birth is usually divided into three sections, although, in practice, it is continuous.

1. First stage labour—opening of the cervix. The start of labour is initiated by hormones which make the uterus contract, causing the fluid-filled placenta to enter the cervical canal. This increased pressure causes the cervix to dilate and the placental sac or water-bag may be forced through into the vagina.
2. Second stage labour—delivery of the animal. The presence of the water-bag in the vagina causes the vagina to dilate, which stimulates the animal to dilate her abdominal muscles, helping the expulsion of the young. The action of the hormone oxytocin on the muscles is responsible for these contractions. At this stage the water-bag ruptures, releasing a large quantity of fluid. The young animal is still enclosed in the inner fluid-filled sac. The contractions increase in frequency and strength and soon the animal is born, often covered in the placental membrane.
3. Third stage labour—expulsion of the placenta. This normally happens quite quickly after the birth is complete. Cows may take one to seven hours to expel the placenta, whilst with sows and ewes it is normally expelled within an hour or so. The suckling of the young stimulates the release of the hormone oxytocin which helps with the expulsion of the placenta. If it is not expelled, which is quite common in cows, the veterinary surgeon may be required to remove it manually before it becomes the source of infection.

The role of the stockperson at parturition is threefold. He or she must:

be able to be patient enough to let nature take its course;
be able to recognize when the circumstances are abnormal;
know when to seek help if the situation dictates.

A great many abnormalities can occur that are concerned with presentation, i.e. the way in which the fetus presents itself in the vagina for birth, but there are several other possibilities. The fetus may be abnormal in some way or it may be too large to pass through the fixed ring of bone, the pelvic circle. This can be a common problem if a large-framed breed of bull is crossed with a small-framed breed of cow. In such cases the stockperson will need veterinary assistance, especially if a caesarian birth has to be performed or, as a last resort, the fetus has to be dismembered whilst inside the mother.

Parturition in the cow

Labour pains are likely to last at least two hours (this time varies widely between animals) before birth takes place. The period can be much longer, especially if the cervical ring is slow to open. The cow may calve either standing or lying down. In normal presentation (Fig. 8.10a) the calf faces the opposite direction to the cow in an upright position with its fore legs fully extended and its head resting on the fore legs. In this position a cow will usually give birth without assistance and it is undesirable to give assistance too soon. The tightest point for the calf is getting its head through the pelvic bones, for the rest of the calf can be squashed, but the bones of the skull will not move. If assistance is given in a straightforward parturition like this, the method is to tie soft ropes onto each of the front legs and to pull in unison with the cow's contractions. The pull must be in a downward direction to avoid the risk of the calf's pelvis becoming locked with that of the mother. This is not as easy as it first appears, because the cow may arch her back, so the direction of pull should be downward in relation to the line of the cow's back. Pulling on the ropes can be done by two people, but for occasions when only one person is available mechanical calving aids can be used. As they exert tremendous force they should only be used by competent, trained stockpeople.

Abnormalities of a forward presentation may vary from something slight, such as the legs being partially retracted when they merely require straightening, to the legs being retained in the uterus or a breech presentation (see Fig. 8.10b, c and d). In all such cases the legs and head must be straightened before birth can take place.

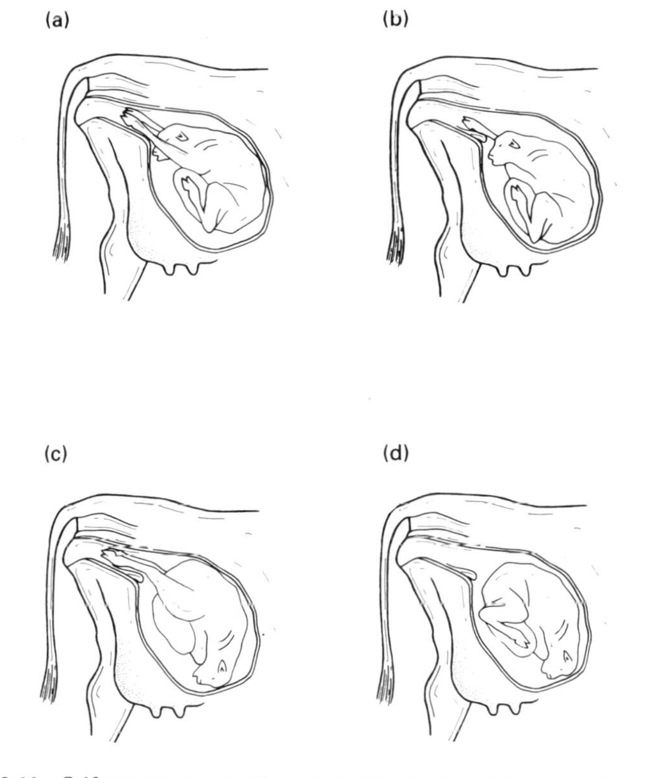

FIG. 8.10 Calf presentation. (a) Normal. (b) One leg back. (c) Backward presentation (hind legs first). (d) Breech presentation.

Occasionally calves are born backwards (Fig. 8.10c). Little difficulty arises if the calf is upright and the hind legs are presented fully stretched, but if either leg is down or if both are down, again, straightening has to take place. When the hind feet can be seen, assistance is usually best, not because the cow is likely to have any difficulty calving, but because of the risk of the calf suffocating. If calving is slow, the calf's head may still be in the fetal fluids within the uterus when blood stops flowing along the umbilical vessels because they are severed or crushed. Absence of gaseous exchange via the placenta and the cooling effect of the air on the calf's wet body may lead to the calf making its first effort to breathe. If more than a very small quantity of liquid enters the respiratory system, the calf is unlikely to survive. With a posterior presentation, therefore, speed is important. If assistance is to be given, adequate manpower must be available and the ropes properly secured before any attempt is made to assist the cow. Pulling should start when the cow makes an effort and should continue until the calf is removed.

The time should be quite short, only a few minutes or so. The calf may need to be suspended upside down, say over a gate, for a couple of minutes to let any fluid drain from its lungs.

When abnormalities occur, calving a cow is a highly skilled operation, firstly in recognizing what is wrong and secondly in rectifying the trouble without inflicting injury on the cow. Experience teaches a great deal, but no stockperson would hope to have a large experience of abnormal calving in his herd. The important thing is to learn to recognize trouble early and to seek assistance at once.

Serious damage can result from applying traction when the abnormality has not been recognized and corrected first. Leg-recognition presents the greatest difficulty. Confusion as to whether the legs are fore or hind legs can easily arise, particularly if the fetus is upside down.

The stockperson must confirm that there are two front legs and a head in the vagina and that they all belong to the same calf. Fore and hind legs can be checked by bending the legs from the hoof upwards: if the first two joints bend the foot the same way, it is a front limb; if the foot bends up at the first joint and the leg bends down at the second joint, it is a hind leg. Note also that there may be twins present, so the number of legs and to which calf they belong must be checked before attaching ropes.

Corrections of abnormalities are not usually easy, but they do become more difficult as time goes on and the fluid has been lost from the uterus, resulting in loss of lubrication.

Parturition in the ewe

Lambing is normally a quicker process than calving, but the same abnormalities may be encountered. Twins are quite normal in sheep and indeed triplets and quads are not at all unusual. The small size of the fetuses usually makes lambing easier, but if assistance is needed the human hand and arm have little room in which to work. Patience, gentleness, a small hand and lots of experience all help to make lambing time a less daunting prospect. Ewes (and lambs) are prone to develop disease as a result of rough handling at lambing, so great care and a hygienic approach must be adopted.

Parturition in the sow

Because piglets are so small in relation to the body size of the sow, farrowing rarely presents any problems. Piglets are normally born head and front legs first, but breech presentation is very common.

Occasionally a large piglet will get caught in the pelvic girdle and will have to be removed by hand, but generally a sow is best left alone.

The length of time for farrowing varies considerably, from three-quarters of an hour to two hours. It is not unknown for sows to give birth to half a dozen piglets, take a break for six to eight hours and give birth to half a dozen more.

Husbandry at the time of parturition

Parturition, although it is a perfectly normal experience, is nevertheless something of a shock to the whole system. It also presents the opportunity of infection to gain access to the uterus and the new-born is susceptible to infection by mouth or navel cord. The opportunity for contracting infection is all the greater because parturition often takes place in the same place as other stock; for example, calving boxes or farrowing pens. The two essentials are comfort and cleanliness.

The cow

Calvings may take place in the field, calving box or some type of loose-box pen. Calving in the field reduces the risk of infection, but if the cow is infected with say, salmonella, it could be spread very quickly over the land. A specialist calving box is the place of choice. Easily cleanable concrete walls and floors are essential. There must be adequate space for the stockperson to pull on calving ropes and the door should open outwards in case a cow lies up against it and refuses to move. As the calf is born, remove any placenta and mucus from its nose and mouth.

If the calf has not taken a breath, tickling its nostril with a piece of straw will make it sneeze, thus helping to expel any fluid. Placing the calf in front of the cow will encourage the cow to lick it; this also stimulates the calf to breathe.

Cows that have just given birth require close observation, as they are prone to a number of illnesses and problems. The ones the stockperson should be on the look out for are:

Milk fever,
Excessive bleeding from uterus or vagina,
Acute mastitis,
Prolapsed uterus,
Retained placenta,
Failure to let down milk,
Infections of uterus and vagina (may develop a few days after calving).

The ewe

As with cows, ewes can lamb inside or outside, the difference being that lambing is usually concentrated into a shorter time period. Thus, if infection enters the flock, it can easily spread. Lambs may need assistance to breathe and once any mucus and placenta have been removed from around the mouth and nostrils, the straw tickling nostril trick, as described for the cow, works well.

The sow

Normally piglets present no problems. They are soon up and on their feet looking for the teats. Sows usually farrow in a specially constructed farrowing crate or farrowing nest, with provision for the piglets to move away from the mother to avoid being laid on. Gilts farrowing for the first time need careful observation as they occasionally savage their piglets. As sows expel their placenta very soon after farrowing, the stockperson must ensure that no piglets are suffocated in the afterbirth.

Because the farrowing crate is used again and again, it needs careful cleaning, disinfection, drying and resting before the next sow moves in.

Colostrum

Farm animals are well developed at birth and usually succeed in standing up in a matter of minutes. Their stance is at first shaky, but greater control develops rapidly and the animal is soon looking for a drink. The first milk is colostrum, a rather sticky liquid much richer in protein than normal milk, but chiefly of interest for its richness in antibodies. These antibodies offer a general protection to the offspring, but they will also be specially related to the disease-causing organisms in the area in which the mother has been living. Colostrum is important, therefore, to all young, but also gives an added protection to those reared on the same farm as the mother.

The digestive system of the young animal is able to absorb the antibodies in the colostrum for a limited time period only. As this property is soon lost (calves can only absorb colostrum in the first eight hours of life), it is essential to ensure that all new-born animals get a good first meal of their mother's colostrum as soon as possible. If lambs or piglets are fostered onto other mothers, care must be taken to ensure that they get colostrum. All colostrum contains antibodies, so a young animal does not necessarily have to have its own mother's colostrum, although it is obviously preferable.

Colostrum is effective even when it has been deep frozen, so a store of colostrum can be built up in case of emergencies. Re-heating of the colostrum must be done carefully, as boiling will destroy the antibodies.

Induction of parturition

Most losses of very young animals occur at, or soon after, birth. If a stockperson could control the time of calving, farrowing or lambing it would be a great asset in reducing losses as he or she could be on hand at the time of the birth. It would also mean that weekend and night-time births could be eliminated. However, in the latter stages of pregnancy the fetus is building up its energy reserves in the liver and muscles ready for the strain of being born. Thus premature induction could lead to weakly animals with no 'will to live', especially if the technique is carried out too early. It should be stressed that natural parturition is best and artificial induction should only be used in special circumstances.

If artificial induction is to be used, then the day of service or insemination *must* be recorded.

Inducing cows

This is not widely practised in the United Kingdom, but New Zealand has had much experience in the technique.

It involves injecting the cow with a synthetic hormone (glucocorticoid) a number of days before the due calving date. The closer to the due date the injection is given, the more likely it is to be a successful live birth.

Large-scale experiments with this technique have shown that treated cows have a higher than average percentage of retained cleansings and a slightly reduced milk yield.

Inducing sows

This is practised quite widely and is usually very successful.

A single injection of prostaglandin causes the corpus luteum to regress, thus reducing the amount of progesterone in the blood. As progesterone is responsible for maintaining pregnancy, the sow will farrow if the level drops.

The majority of sows can be expected to farrow twenty-four hours, plus or minus six hours, after the injection. If the treatment is too premature, a lot of weak piglets may be born. It is advisable to calculate the average gestation period for the herd, as this may vary from the average by a day or so. It should be noted that prostaglandin must be administered under the supervision of a veterinary surgeon.

Inducing ewes

Control of the date of lambing is usually achieved by means of synchronized oestrus at the start of the gestation period. As with cows, an injection of a synthetic hormone can be administered to the ewe so that she will lamb about two days after treatment. This may be useful in some individual cases, but expense probably precludes it from becoming regular practice in the commercial flock.

Chapter 9

The udder

The udder, or more correctly the mammary glands, are specialized skin glands that are held on the outside of the body. The udder of a cow consists of four mammary glands, a pig has twelve to fourteen, and a sheep has two. Strangely, each gland of a pig's udder or sheep's udder is usually referred to as 'quarter' as is usual with cattle.

The function of the udder is to synthesize milk from blood nutrients in the specialist milk-producing cells deep in the udder tissue.

Structure of the udder

The udder is supported by connective tissue, although the cow's udder has a sheet of suspensory ligaments down the centre of the udder (from back to front) and down both sides. These ligaments can stretch with age, giving the characteristic pendulous udder that older cows often display. Stretching of the udder is also caused by the udder filling with milk, the stretching allowing for about 50 per cent more milk to be accommodated.

The teat opening is held closed by a sphincter muscle which leads into the teat cistern (Fig. 9.1). There are folds of tissue which help hold milk in the gland cistern which is situated above the teat cistern. Leading into the gland cistern are large ducts and spaces known as sinuses. The function of the cisterns, ducts and sinuses is purely for storage of milk as they are not milk-producing structures.

The large ducts branch into smaller ducts, ending up in a microscopic lobule of milk-producing alveolar cells. These cells are surrounded by blood capillaries from where the nutrients are obtained to synthesize the milk.

Milk let-down

Touch-sensitive nerves surround the end of the teat and nerves that sense temperature and pressure are situated on the outside of the udder. In conjunction with these, nerves running from ears, nose and eyes relay information to the brain that milking is about to take place. This stimulates the pituitary gland to release the hormone oxytocin which has the effect of squeezing milk from the milk-secreting tissue into the milk ducts and milk cistern. Pressure in the udder rises due to milk finding its way into

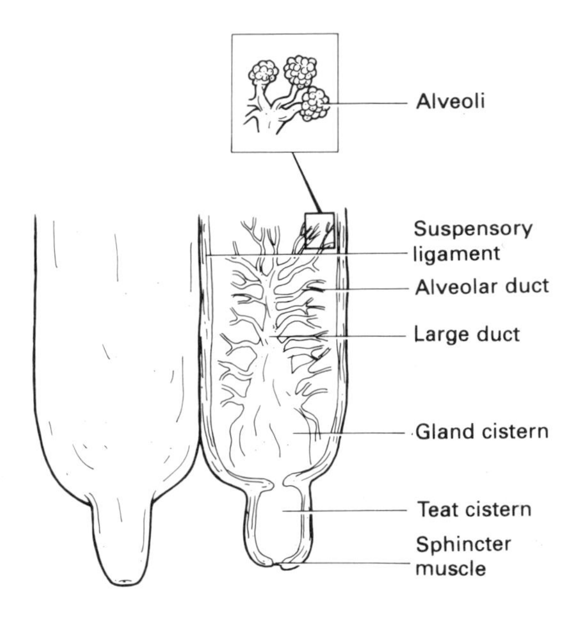

FIG. 9.1 The structure of a cow's udder.

the cistern. The sphincter muscle at the end of the teat holds back the pressure until it is stimulated to open by suckling action or the milking machine.

Unfortunately this nervous/hormonal response can work in reverse. If the animal being milked experiences any changes in the normal milking routine such as pain in the udder, strangers, or unsympathetic handling during milking, the size of the blood vessels will decrease and prevent oxytocin reaching the milk-secreting tissue. If the animal is frightened, adrenalin (from the adrenal gland) is secreted which also causes constriction of blood vessels and so blocks the action of the oxytocin.

For dairy cows that are milked through a parlour, not only has the removal of milk from the udder got to be effective, it also has to be reasonably quick, about five to six minutes for an average cow. It is essential, therefore, to have a well-planned milking routine to avoid any unnecessary delay.

Lactation

The milk-secreting cells draw the raw materials for milk synthesis directly from the capillaries surrounding the alveoli. It is estimated that in cows about 500 litres of blood pass through the udder in order to produce one litre of milk.

Pressure within the udder is low after a milking, but soon begins to increase as milk is produced. Milk is secreted from the cells at an inverse rate to pressure within the udder, so that as the pressure builds the secretion of milk slows down. This is the reason why dairy cows that are milked three times a day produce more milk than cows

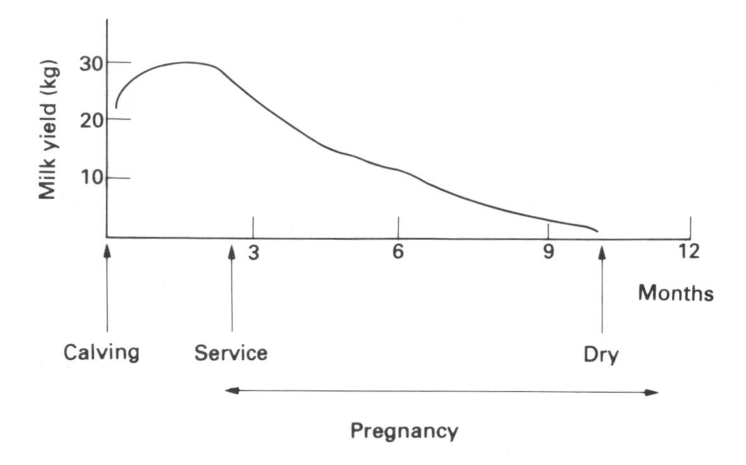

FIG. 9.2 Typical lactation curve for a dairy cow.

on twice a day milking. The milk fills the alveoli and pressure causes it to flow down to fill the udder spaces and cisterns.

The udder of the animal breeding for the first time enlarges in later pregnancy and begins to secrete milk as parturition approaches. The first milk of each lactation is colostrum, but this changes rapidly and milk is of normal composition after about three days. Provided that milk is regularly removed by milking or suckling, the quantity increases for a period—up to about six weeks in the cow, about three weeks in the sow and ewe—and then tends to decrease. Any quarter that is not milked becomes swollen with milk for two to three days, the milk begins to be re-absorbed and the quarter contracts and dries up. This can easily be seen when a sow has fewer piglets than number of teats, as each piglet suckles on its own teat. Sows will produce milk for a long time if the litter is allowed to suckle, but lactation ends when the litter is weaned. Lambs are normally artificially weaned after about four months on their mother's milk.

The cow, on the other hand, is different. Dairy cows have been bred to produce more milk than the calf requires and so will keep on producing milk in small amounts of up to three years if milked continually. However, modern dairy cows start another pregnancy nine to twelve weeks after calving and, as the growing embryo demands more nutrients the yield falls, producing a typical lactation curve (Fig. 9.2) of peak yield at about six weeks after calving with production slowly tailing off as pregnancy progresses.

Cows are normally dried off six to eight weeks before calving in order to give both cow and udder a rest. Milk-secreting cells are lost during lactation, so one of the functions of the dry period is to regenerate these cells.

Factors affecting the milk yield of dairy cows

There are a number of factors that will affect the milk yield, some environmental, some management-related and some genetic.

It is worth listing these briefly so as to appreciate the complexity of achieving maximum yield from a dairy herd.

1. Feeding. The most important factor governing the yield is the quality and quantity of the food available. Fresh grass in the spring and summer will lead to increased yields.
2. Breed. There are variations in yield owing to breed; for example, the average milk-yield for a lactation of a British Holstein dairy cow would be about 6500 litres, whilst that of a Jersey would be about 4000 litres.
3. Age. Up to about the fourth lactation the average yield will increase, but after about the seventh the yield will fall. This will vary between individuals.
4. Health. It goes without saying that healthy cows will yield more than unhealthy cows. A long period of illness may permanently affect the amount of milk a cow will yield.
5. Water. Clean drinking water is essential to maintain yield; this is especially important during the summer months.
6. Temperature. Extreme heat or cold will cause a reduction in yield.

Part 2. Disease and Disease Prevention

Chapter 10

Causal organisms of disease, immunity and the nature of disease

The cause of disease

Disease can be defined as any adverse deviation from the normal. Often two or more factors contribute towards a disease, thus there is more than one way in which to try to prevent or cure a disease.

Bacterial invasion of the body may result in disease, as may invasion by viruses and parasites, but invasion of the body is not necessarily followed by disease. Other factors such as the environment may be more important either as an actual cause or as a deciding factor in the severity of a disease. Other causes could be a deficiency of some essential item in the food, under-nutrition, poison, injury or functional failure.

Causal organisms

Bacteria

Bacteria are usually described as one of the simplest forms of life. They can be clearly recognized under a high-power microscope and have varying shapes, mostly appearing as spheres, known as cocci, or rods. Some idea of the size is given by the fact that the anthrax rod, one of the largest bacteria, is about 0.005 mm in length.

Bacteria consist of a single cell with a thick cell wall enclosing the cytoplasm and the genetic material. Bacteria do not have a specific nucleus, as have animal and plant cells, but they do contain the genetic information required for multiplication. Nutrients are absorbed through the cell wall and waste products are excreted from the cell. These waste excretory products are often the cause of illness in an animal.

Bacteria multiply by splitting in two. Under favourable conditions this splitting can take place every half hour—a rate which would give one bacterium 17,000,000 descendants in twelve hours! In practice, of course, bacteria live in a competitive world and rarely find such ideal conditions.

Some pathogenic bacteria have the power to produce spores. The spore is a resting form in which the bacterium can exist in a state of suspended animation, resisting adverse conditions of drought, heat and cold. Vegetative (non-sporing) bacteria will usually be killed by pasteurization, that is being held at 63°C for half an hour in a liquid medium, whereas short boiling will fail to destroy many spores. The anthrax spore, for instance, requires boiling for at least ten minutes and can exist in soil for up to forty years.

Viruses

Viruses are much smaller than bacteria and can only be seen using an electron microscope. They consist of a core of nucleic acid (DNA or RNA) surrounded by protein. Viruses cannot carry out the normal functions of growth and reproduction on their own and so need to invade other cells in order to do this. They, in effect, 'hijack' the cell and use it for their own purposes. The virus multiplies within the cell until the cell is full of viral particles. It then bursts, releasing more virus to infect surrounding cells.

Because of their specialized nature, viruses cannot live outside the host's body and have to be transferred from animal to animal in some medium; for example, the foot and mouth virus can be transferred by milk from an infected animal.

Viruses (and bacteria) are often specific to the site where they cause infection. Some will only grow in the gut, while others choose the lungs or nasal passages.

Mycoplasms

Mycoplasms are smaller organisms than bacteria but larger than viruses. They are responsible for a type of mastitis in dairy cows and can cause lameness in pigs.

Fungi

Fungi from soil or mouldy hay or straw are known to cause abortions in cattle and pigs—mycotic abortions—and to cause ringworm in cattle and domestic pets. Fungi will multiply quickly under favourable conditions, usually a damp, warm environment and, as they form spores, are quite resistant to disinfection.

Parasites

These can be anything from the minute single-celled protozoans, for example coccidia of the intestine that cause problems in pigs and poultry, to the large 'animal' parasites such as lice and worms.

Parasites live either all or part of their lives on or in their host. Some have a complex life cycle that entails leaving the primary host to go to a secondary host. Understanding this life cycle is essential in order to find a method of control or prevention.

Parasites are dealt with in two separate chapters—Chapter 22, External Parasites, and Chapter 23, Internal Parasites.

Immunity

The mechanism by which animals resist disease is by use of the immune system which produces antibodies to fight infections. In order to understand how these antibodies are produced and acquired it is best to begin with the development of immunity in the new-born animal.

A new-born animal is free from most disease organisms but is able to combat disease it meets at birth because it has acquired 'passive' immunity to some diseases from its mother. If the mother has been exposed to or vaccinated against a disease, her immune system will have produced antibodies to fight that disease. Those antibodies will be in her blood stream and so will be passed to the fetus in the womb. The young animal will also acquire antibodies from the colostrum. These antibodies will protect the animal for about three to four weeks, depending on how much immunity is received via the colostrum.

When the young animal is exposed to a disease it will develop 'active' immunity. That is, the animal's own immune system will produce antibodies to fight the disease it has been exposed to. Active immunity can also be brought about by vaccination. This stimulates the production of antibodies to a particular disease as if the animal had been exposed to that disease.

An animal takes some time, perhaps two to three weeks, to build up active immunity against a disease and during this time in young animals the passive immunity is waning. This leaves a 'window' when the young animal is most susceptible to disease (Fig. 10.1). So, in practice, it is essential to make sure that any challenge the animal has to a disease-producing organism is gradual and does not overwhelm the immune system.

Vaccines

A vaccination will give an animal a low dose of a disease organism which will not cause symptoms of the disease but will stimulate the immune system to produce antibodies and so provide active immunity.

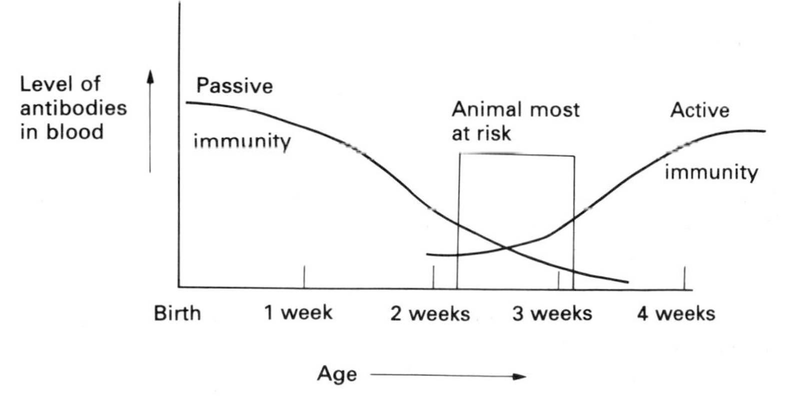

FIG. 10.1 'Window' when the young animal is most susceptible to disease.

Vaccines are used when there is a history of a specific infection on the farm. They may be produced by altering the infectious organism in some way, e.g. by exposure to X-rays, so reducing its virulence; the so-called 'live' vaccines. When injected, the animal's body produces the antibodies to fight that disease. Subsequent infection will activate the immune memory of the animal so it produces the right antibodies quickly and in large amounts. 'Dead' vaccines using killed organisms may also be used. Live vaccines usually produce stronger active immunity than do dead vaccines, although they may not last any longer.

Occasionally a vaccination may not be effective because it conflicts with the animal's passive immunity to a disease. Thus the timing of vaccinations is important and this is usually referred to in the leaflet of instructions that comes with the product.

The nature of disease

Disease may be either a structural defect or a functional one. The structural defect is clear when a bone is broken or an abscess is formed. A functional defect denotes that the part is apparently normal but is not carrying out its normal function, e.g. a paralysed muscle which causes a limb to be limp or a muscle in spasm which causes a joint to be severely flexed.

Acute and chronic disease

These words have a clear meaning in medicine rather different from that in ordinary usage. They are contrasting terms. 'Acute' denotes a disease that is quick in onset and short-lived, whilst 'chronic' denotes a slow, longer-lasting condition. The words are used comparatively and therefore denote no fixed time. A disease like liver fluke in sheep, which normally lasts for months even when fatal, will be described as acute when a sheep wastes away rapidly in a few weeks, but a normally acute disease like swine erysipelas which can kill in a few days will be described as chronic if it settles in the heart and becomes protracted to a few weeks. The word 'acute' does not denote a severe disease, though many acute diseases can be severe. Acute swine erysipelas, for instance, though often fatal if untreated, responds well to treatment, whereas the chronic form of the disease does not respond and is usually fatal.

The word subacute is used to describe a condition between acute and chronic, while a disease which kills very quickly is call peracute.

Infections and contagion

A disease is infectious if it can be transmitted from one animal to another without direct contact. This usually means that the disease is caused by either a bacterium or a virus, which is carried in the air and breathed or swallowed by the animal. A contagious disease is one which is transmitted by contact. The best examples are external parasites which cannot fly, such as mites and lice.

The words 'non-infectious' and 'non-contagious' have important and obvious meanings. Deficiency diseases such as milk fever are both non-infectious and non-contagious, even though a whole group of animals could be affected at the same time.

Epizootic; enzootic; sporadic

The word 'epidemic' denotes a disease spreading amongst people. The same word is sometimes used about animal disease, but purists insist that 'epizootic' should be used to denote a spreading disease in animals.

Similarly, an enzootic disease is a disease continuously present amongst a group of animals without spreading fast, as opposed to an epizootic disease which runs its course in the group and then tends to die away.

A sporadic disease is one which suddenly appears, affecting one or more animals, but then as suddenly disappears.

Bacteraemia; septicaemia; toxaemia

When bacteria gain entry to the animal body they tend to be localized and the defence mechanism tends to keep them that way. If, however, bacteria gain entry to the blood a state of bacteraemia occurs in which bacteria use the animal's circulatory system to get carried all over the body.

When the bacteria are actually multiplying in the blood the condition is described as septicaemia. This is a very serious development; unless the animal rapidly begins to fight the disease, the result is likely to be fatal.

Sepsis is the condition in which bacteria are active in a part of the body with resultant breakdown of tissue. The classic example is the septic wound, usually a much less serious matter than septicaemia.

The word 'toxin' is used mainly to describe the poisons in the animal body derived from bacteria and other invading organisms, or the breakdown products of the animal's own tissues. When such products gain access to the blood a state of toxaemia is produced. The importance of this state depends on a number of factors, chiefly the potency of the toxin, as some can be fatal in minute quantities.

Fever

Any condition in which the temperature is raised can be termed a fever, and diseases which lead to fever are termed febrile in contrast to non-febrile or afebrile diseases in which a rise of temperature does not occur. Fever is usually the animal's response to the invasion of bacteria or other micro-organisms or to a state of toxaemia. A moderate rise in temperature is not essentially a bad thing, for it is part of the animal's mechanism for combating disease.

Lesions

A lesion is tissue damage resulting from disease. Fundamentally, lesions are of two main types, inflammation and tumour formation. Inflammation is by far the more common, especially in farm animals which are rarely allowed to live to old age. Tumours are largely confined to older individuals.

Inflammation

The word implies heat and in acute inflammation the affected part is always hot. The classic description of acute inflammation lists swelling, heat, pain and reddening, to which may be added impairment of function. All these signs are due to the increased flow of blood to the affected part.

Inflammation is the normal reaction of healthy tissue to any form of injury or irritant. The word 'irritant' covers not only physical and chemical agents but also the effects of living organisms, as in an abscess. Though painful, inflammation is the first essential in the animal's recovery. Only by bringing up defences in force can the effects of the irritant be overcome.

Chronic inflammation is the reaction of tissue to a persistent low-grade irritant. Blood to the part is not greatly increased, so the signs associated with acute inflammation are not present. Over a period the affected part will tend to become enlarged due to the laying down of fibrous tissue or, if bone is involved, thickening of the bone. Such lesions are most likely to occur in the limbs and may interfere with function or cause pain if they involve pressure on other tissues.

Tumours

Tumour means literally any swelling, but has come to have a special meaning. It is the word popularly used to mean new growth (neoplasm), a swelling without inflammation, due to the development of tissue serving no useful purpose to the animal. Tumours are of two clearly defined types, benign and malignant.

The word 'benign' is used in a comparative sense as a contrast to the much more dangerous malignant tumour. Benign tumours are usually contained within a capsule. Growth is usually slow and the lump may remain static in size for long periods. They often do negligible harm in themselves and if the swelling is external they may be only an unsightly blemish. Internally, however, they can be dangerous by imposing pressure on a vital organ by constriction of a passage, the intestines for instance. Such tumours can usually be surgically removed, though when they are causing no trouble it may be advisable not to interfere.

A malignant tumour has no capsule and grows by invading and destroying the surrounding tissue. Starting from a small beginning the rate of growth tends to accelerate. When they penetrate and destroy normal tissue they must inevitably interfere with the circulatory system, at such times, a few tumour cells may break away and be carried in the blood stream to start another tumour when they settle elsewhere in the body. Surgical removal of the original tumour may therefore only give temporary improvement, because daughter tumours (metastases) may arise in various other parts of the body.

Chapter 11

Notifiable diseases

In order to control the spread of serious diseases, or to stop their entry into the United Kingdom, the Ministry of Agriculture has legal powers under the Animal Health Act, 1981 (which incorporates the Diseases of Animals Act, 1950). Under this Act there are various Orders that make certain diseases 'Notifiable', that is, if there is an outbreak or suspected outbreak of one of the listed diseases, it must be reported to the police or the Ministry of Agriculture. However, the owners of animals have other responsibilities under the Act, responsibilities which are often delegated to the stockperson. These include:

To report any cases of abortion in cattle.
To keep a record of all animal movements on to or off the farm.
To segregate diseased animals.
To report all cases of sudden death in cattle.
To provide facilities and assistance to persons inspecting diseased animals.
To prevent the sale of diseased or suspected diseased animals in livestock markets.
To stop diseased animals from straying onto roads.
To dip sheep at certain times of the year.

Associated with the legislation on notifiable diseases, there are some other pieces of legislation of which farmers and stockpeople should be aware. For example, The Waste Food Order, 1973, lays down various regulations concerning the collecting, processing and feeding of swill. It is aimed primarily at pig farmers in order to stop the spread of foot and mouth disease, swine vesicular disease and swine fever. There are various Zoonoses Orders (1988 and 1989) that require some zoonotic diseases (that is, diseases which are transmissible from animals to man, see Chapter 24) to be reported to the Ministry of Agriculture. At the moment those included are salmonellosis, brucellosis and bovine spongiform encephalopathy (BSE).

The reader should note that these are *some* examples of current legislation. There are many other regulations in force, too many to list in this book. Suffice to say that anyone who has responsibility for livestock must be aware of the legal requirements appropriate to that class of stock.

Notifiable diseases

Notifiable diseases have some form of legislation to control them or to eradicate them from the United Kingdom. Many are very infectious, causing severe economic loss to the farmer, e.g. foot and mouth, others are dangerous to both animals and man, e.g. anthrax, whilst others may cause disease in humans if contaminated products are consumed, e.g. tuberculosis.

Table 11.1 lists some important notifiable diseases. There are others, but, because they have not been seen in the United Kingdom (or are unlikely to be seen) for some years, they have not been included in this book.

TABLE 11.1 *Some important and notifiable diseases.*

Anthrax
Aujeszky's disease
Bovine spongiform encephalopathy (BSE)
Bovine tuberculosis (TB)
Brucellosis
Enzootic bovine leucosis (EBL)
Foot and mouth disease
Rabies
Sheep scab
Swine fever
Swine vesicular disease (SVD)
Warble fly (Warbles)

Anthrax

Anthrax is caused by the bacterium *Bacillus anthracis*. It affects cattle, sheep, pigs, horses and many other animals both wild and domestic. Anthrax can also affect humans, although most cases are related to occupation, for example people working in the meat trade or working with bone meal.

Most animals that become infected die quite quickly (within hours) with no noticeable symptoms having been seen. Some may live for a few days, having a high temperature (41–42°C in the case of cattle) and passing small amounts of blood-stained diarrhoea. Pigs can become ill by eating infected meat. They develop a swelling around the throat and neck and have difficulty breathing.

The spores of the anthrax bacterium have the ability to remain dormant in the soil for many years. When permanent pasture or long-term leys are ploughed up, some spores may be brought to the surface to become active again. This is the reason why grazing animals are most at risk from the disease.

If the disease is confirmed, the animals will be slaughtered and their carcasses, bedding and faeces will be burnt on the farm by the Ministry.

It is important that any cases of sudden death in cattle are reported to a veterinary surgeon who will carry out a rapid blood test to see if the cause is anthrax. On no account should the carcass be cut into or handled in any way until the possibility of anthrax has been ruled out.

Aujeszky's disease

Aujeszky's disease is a virus disease which occurs mainly in pigs, but it can affect cattle, goats, dogs and cats. It is also known as pseudorabies as it causes similar symptoms to true rabies in cattle.

In pigs all age groups can be affected, the symptoms being trembling, convulsions, coughing and other symptoms associated with the central nervous system. Pregnant sows may abort or have stillborn or mummified piglets. Very young piglets may die. As these symptoms are the same as many other pig diseases, diagnosis is by blood test.

There is a slaughter policy in operation for Aujeszky's disease. This may be for selected animals or for the whole herd depending on the severity of the disease.

In 1983 U.K. pig producers funded, by means of a levy on all pig carcasses, an Aujeszky's eradication policy. In 1989 the United Kingdom was declared free of the disease.

In cattle the symptoms are rubbing or scratching as if to stop a severe itch. The animal usually dies within two days.

Bovine spongiform encephalopathy (BSE)

At the time of writing, the cause of BSE is unknown, although there is much conjecture. The disease is similar to scrapie in sheep (*see* page 167), suggesting that BSE crossed from sheep to cattle, probably via meat and bone meal from sheep being incorporated into cattle feeds. This practice was banned in July 1988.

The symptoms in cattle are nervousness, increased sensitivity, abnormal gait, and slipping and falling over. Death follows up to six months later. It affects cattle between three and eleven years of age, but most cases occur between three and five years of age.

At the moment there is no evidence that it can be transmitted from cow to cow or from dam to offspring. This suggests that the cow is a 'dead end host', i.e. the disease does not pass to other species, but to date this is not proven either way.

Legislation

Animals suspected of being infected with BSE are subject to a slaughter-with-compensation policy, the compensation being 100 per cent of the market value. The brain is removed for examination and the carcass incinerated.

Milk from suspected animals must not be used for human consumption.

To prevent transmission to European countries, the European Community has banned the export of live cattle born before 18 July 1988 or born to females known to have been, or suspected of, having been affected by BSE.

To stop the disease getting into the human food chain, food manufacturers are not allowed to use the brain, spinal cord, spleen, tonsils and intestines of cattle as part of any food product.

The future

As with any 'new' disease, time will bring a better understanding of its ultimate effects. In the meantime, the United Kingdom Government has targeted over

£12 million (1990) over three years for research into the cause and transmissibility of the disease.

Bovine tuberculosis (TB)

Tuberculosis is a bacterial disease that can affect many animals (pigs, sheep, goats) including man, but is usually referred to in relation to cattle. In the 1920s many cattle were infected with the disease and developed tuberculosis in the udder. This resulted in infected milk which passed tuberculosis on to humans.

The fight against tuberculosis started in 1915 with the Tuberculosis Order, designed to detect clinical cases. The real breakthrough came in 1935 with the start of the voluntary Attested Herds Scheme which, by 1950, provided enough 'clean' cattle to start compulsory eradication. By October 1960 the whole of the United Kingdom was virtually clear of Bovine TB.

Cattle are tested by means of a comparative skin test. This involves two injections into the skin, one to test for avine TB, the other for bovine TB.

Two nodules form at the injection site and the size of the nodules is measured. A large bovine TB nodule and a small avine TB nodule indicates that the animal has been exposed to bovine TB. Further tests are done and the animal is slaughtered if found to be a 'reactor'.

Symptoms of bovine TB are rarely seen, but include a short, dry cough, indicating TB of the lungs. A thickening of the udder tissue indicates the more important TB of the udder.

Brucellosis

Brucellosis is a bacterial infection that can affect most animals, including man, but is most often associated with cattle, the causal organism being *Brucella abortus*. It is sometimes known as 'contagious abortion', but as it can affect bulls as well as cows, this is a misnomer.

The bacteria live in the udder, uterus and testicles and the disease affects animals of breeding age. Cattle are infected through grazing on contaminated pasture, licking aborted calves or being served by an infected bull.

Infection may result in abortion, which can occur without previous symptoms, although some cows will carry their calves for the full pregnancy. If the cow does go full term, the placenta is often retained and the cow has a reddish-brown discharge which may last for fourteen days. Bulls show no obvious symptoms whatsoever. The disease can spread very quickly through a clean herd, resulting in an abortion 'storm' where most of the herd aborts. The result is a huge loss of calves and milk with resulting financial losses.

Eradication of the disease began in 1971 and by 1983 the United Kingdom was virtually free from the disease. This was brought about by the compulsory use of a vaccine (S.19) on all calves between three and six months of age. This reduced the incidence of abortions and so stopped the spread of infection within herds. In addition, cattle were regularly blood tested and milk samples from each farm were checked.

Vaccination stopped in 1979, so the national herd is now highly susceptible to the disease.

Legislation

Farmers and stockpersons have a legal duty to report all cases of abortion in cattle.

It is also illegal to sell a cow that has aborted until at least two months after the abortion.

Enzootic bovine leucosis (EBL)

Enzootic bovine leucosis is a virus disease of cattle that produces a marked swelling in the lymph nodes. These growths may occur on the neck, causing the animal to breathe noisily, but they can appear at any of the lymphatic sites. Other symptoms are severe weight loss and anaemia. There is no treatment and animals with the disease slowly die. As other diseases can cause enlarged lymph nodes, diagnosis is by blood test.

The disease is passed on mainly via the colostrum, but infected blood can also be responsible for some cases of transmission. This may occur from dirty and contaminated injection equipment or possibly from blood-sucking insects.

In 1982 the Ministry of Agriculture introduced the Enzootic Bovine Leucosis Attested Herds Scheme as a first step to eradication of the disease in the United Kingdom.

Foot and mouth disease

Foot and mouth is a virus disease that affects cattle, pigs, sheep and goats. Other animals, for example birds and hedgehogs, may be involved in its transmission.

The first symptoms to be seen are dull, lethargic animals that are off their food. They look tender on their feet, shifting their weight from one to another and they may be drooling from the mouth. The animals may have a high temperature and milking cows show a reduction in yield. When examined closely, blisters can be seen on the tongue and between the claws of the feet.

The virus is present in the fluid-filled blisters that eventually burst and in the saliva, urine, faeces and milk. It can also be present in or on anything that has been in contact with the animals, for example, straw, bedding, sacks, hides, hair, vehicles and stockpeople. The wind was thought to be responsible for much of the spread during the outbreak of 1967–8. It is a highly infectious disease that spreads easily between animals and farms.

The disease is only fatal in about 5 per cent of cases (higher in very young animals), but it does cause severe economic loss in terms of reduced weight gain and milk yield.

Legislation

In order to keep possible sources of infection to a minimum, there are strict controls on the import of live animals into the United Kingdom. Also the importing of meat is strictly controlled as the virus can live for some time in the carcass. There are regulations controlling the collection, storage and feeding of swill to pigs, as infected meat could pass the disease back to the animals. Swill has to be boiled in an approved (licensed) boiler for a controlled length of time before it can be used.

Control of an outbreak is based on a slaughter policy. Infected herds are slaughtered and the movement of all animals within a ten-mile radius of infected farms is controlled. Slaughtered animals and infected materials are either buried or burnt. The farm is disinfected and cannot be restocked until a specified period of time has elapsed. (This time period is variable according to the severity of the outbreak in the area, but typically it may last from four to six weeks.)

Some countries use a vaccine to control the disease, but as this does not confer full immunity and some animals can become carriers, the slaughter policy remains in the United Kingdom. However, following the 1967–8 outbreak, supplies of the vaccine have been stockpiled in case an outbreak should get out of control.

Rabies

Rabies is a virus disease that affects nearly all mammals, including man. The virus attacks the central nervous system, resulting in deranged behaviour, paralysis and, almost always, death.

There has not been a serious outbreak in the United Kingdom since 1921, although several cases have occurred inside quarantine areas.

Such is the seriousness of rabies that a whole host of regulations exist, firstly to stop the disease entering this country and secondly to deal with any outbreak that may occur. The main defence against the disease is The Rabies (Importation of Mammals) Order, 1971, which states that mammals being imported into this country must undergo six months of quarantine upon entry.

Sheep scab

Sheep scab is caused by two mange mites (for diagram of mite refer to Fig. 22.3, page 171):

1. *Psoroptes communis*.
2. *Sarcoptes scabei*.

Both are covered by the Sheep Scab Order, 1977, although their symptoms are slightly different.

Psoroptic mites attack the skin of the wool-bearing areas, whereas the sarcoptic mite affects the non-woolly areas of the head, ears and around the eyes.

The mite bites into the skin, causing a small wound. It then feeds on the fluids that ooze from the wound, causing intense irritation to the sheep. The affected animal will rub itself against fences, gate posts, trees and similar objects, causing the wool to become damaged and detached. The skin becomes thickened and pus-filled sores begin to form. Secondary bacterial infection of these sores is quite common. A number of sheep may die and many will quickly lose their body condition.

The disease is highly contagious, it may cause severe suffering to the sheep, and may result in huge economic losses for the farmer.

The life cycle of the mite is completed in about fourteen days, with the mite remaining on the sheep the whole time. The mites are active during the winter months (when the symptoms are most apparent) but remain on the sheep during the summer.

Legislation

It is mandatory to dip sheep using an approved dip in order to prevent the disease. Until 1988 two dippings per year were the legal requirement. In 1989, unless a farmer suspected sheep scab or it was confirmed in the flock, only one dip in the autumn was necessary. This may change in the light of experience, i.e. depending on the number of cases that continue to appear.

Swine fever

Swine fever is a highly contagious virus disease of pigs. It may affect all ages of pigs in the herd, causing abortion in pregnant sows, vomiting and loss of appetite in growing pigs, diarrhoea, purple skin and a discharge from the eyelids.

It may exist in a subclinical form, resulting in mummified piglets being born, whilst other live piglets may have muscular tremors.

The virus can be transmitted by carrier sows or by contact with infected faeces. These faeces may originate from vehicles used for transport or from livestock markets.

The virus can also be carried in pig meat products, especially those that originate from countries where swine fever is endemic. As the virus is easily killed by heating, it is important that all swill food is heated according to the specific regulations.

Legislation

Herds affected by swine fever are subject to compulsory slaughter. The disease was officially eradicated in 1966, but occasional outbreaks continue to occur.

Swine vesicular disease (SVD)

This virus disease first appeared in the United Kingdom in 1972. At first it was thought to be another outbreak of foot and mouth disease (*see* page 81), as the symptoms are very similar, but tests soon confirmed that SVD is caused by a different virus.

The symptoms are lameness, with blisters appearing in the space between the claws of the foot, with others on the snout, mouth, tongue and sometimes the teats.

It is spread by pigs coming into contact with carrier animals or contaminated faeces (maybe left on vehicles), or by being fed contaminated waste food.

Legislation

The Movement and Sale of Pigs Order, 1975, is designed to allow pigs that have been moved to be traced back to their original owners. It means that anyone wanting to move pigs must obtain a licence to do so. This applies to all pigs except those going directly to a slaughterhouse, or slaughter market, from a farm that does not use swill feed.

SVD affected herds are subject to compulsory slaughter.

Warble fly (warbles)

The maggots of warble flies cause small swellings, about the size of a marble, on the backs of cattle during spring and early summer. These maggots are results of 'attacks' by adult warble flies during the previous summer.

FIG. 11.1 A dairy cow being treated with a 'pour-on' compound. The active ingredient is absorbed through the skin, into the blood stream and so is distributed throughout the body.

There are two types of warble fly, *Hypoderma bovis* and *Hypoderma lineata*. The adult flies lay their eggs on the skin of the animal's legs and abdomen during the summer months. The fly has a particular buzzing sound which, when it nears cattle to lay eggs, causes the cattle to rush madly about the field, sometimes causing injury to themselves. The eggs hatch into larvae which burrow into the skin and move into the tissue of the legs. The larvae then migrate upwards, growing in size all the time until about mid-January when they reach the animal's back. When under the skin of the back the larvae puncture the hide to make a breathing hole. They stay under the skin, forming a pus-filled sac around themselves, until April or May when they can be seen as quite large lumps under the skin. After a period of time they emerge as large white maggots (larvae) that drop to the ground to pupate. Between four and six weeks later the pupae develop into adult flies to start the cycle once more.

The damage done by warble flies can be quite considerable. In bad cases the animal may lose body condition and the abscesses (containing the maggot) near the spine can occasionally lead to paralysis or death. Also, carcasses can be ruined because the larva leaves tracks where it travels through the meat.

Treatment is by use of a 'pour-on' organophosphorus compound (see Fig. 11.1), but only at certain times of the year. It should be not be used between 20 November and 15 March, as larvae may be killed when they are near to the spine. This could result in a reaction, causing pressure on the spine and possible paralysis.

It should be noted that ivermectin (effective against warbles, lice, mange, intestinal worms and lungworms) can be used at any time of the year.

Legislation

Farmers and stockpersons have a legal obligation to report all suspected cases of warble fly to the appropriate authorities. Also, if cattle are treated for warbles in the spring, then any cattle of twelve weeks and over must be treated in the autumn.

Chapter 12

The prevention of disease

There is no single factor that will prevent disease coming on to or spreading within a farm. Disease usually occurs because a number of factors are working against the animal, for example the weather, its nutrition or the number of stock in one building. Therefore, the stockperson has to bear in mind that there is more than just one way of preventing disease.

All livestock units should formulate a comprehensive policy to prevent disease. Whilst all sick animals must be treated, the aim should be to prevent animals from contracting illness in the first place. This chapter will deal in very general terms with the factors that ought to be considered when formulating a disease prevention policy.

To help simplify this process, the prevention of disease can be divided into two areas:

1. Prevention of disease coming on to a livestock unit.
2. Prevention of disease spreading within a livestock unit.

Prevention of disease coming on to a livestock unit

Health of bought-in stock

Any animals brought on to the unit must be healthy. In the case of newly purchased animals they should come from a reputable source, preferably with some form of background history. All new stock should be inspected to ensure they are healthy. It is worth noting some general signs of health that are applicable to most animals:

Bright alert eyes.
Clean tail.
Licked, groomed coat (if applicable).
Navel not swollen, hot or hard.
Even breathing.
No sign of discharge from the nose.
No coughing.
Healthy appetite.

In the case of pigs, these should be purchased from one source only, preferably the same farm. This ensures that all the stock will have antibodies against the infections to which they will be exposed when moved on to the new farm.

Some livestock units (mainly pig units) maintain a closed herd policy. This means that no replacement stock are bought in from an outside herd. All stock is bred on the unit using artificial insemination or embryo transfer to provide new blood and so avoid inter-breeding, whilst still maintaining genetic improvement.

Quarantine

Livestock units should have somewhere to house new stock that is away from, or separate from, the main unit. There are three main reasons for this:

1. To ensure that the new, possibly diseased animal, does not come into contact with the general farm population.
2. If the new animal is diseased, some symptoms may show themselves whilst it is in quarantine.
3. To expose the new animal to a mild dose of diseases that are prevalent on the new unit in order to build up active immunity against those diseases.

Active immunity is usually achieved by introducing some dung or bedding from the main unit into the quarantine pens. However, the level of exposure is critical. If the new animal is overwhelmed with new pathogens, it may contract a disease; too little exposure and a satisfactory level of immunity may not be achieved.

The normal time period that the new animal would spend in quarantine is about three weeks.

If new animals are to be housed in a building that has been cleaned and disinfected, as in the case of bought-in calves, then no quarantine is necessary, as they will gradually build up their active immunity as they get older and move through the system.

Vaccination

All new stock must be vaccinated against any known on-farm diseases as part of the normal farm vaccination programme. Examples of these would be salmonella vaccinations for calves, rhinitis vaccinations for pigs, and vaccination against clostridial diseases for sheep.

It should be noted that some vaccination programmes begin with two vaccinations administered in a short time period followed by six-monthly or annual booster doses.

Isolation of the unit

Some intensive pig and poultry units are quite literally 'off limits' to anyone except the personnel that work there. They are surrounded by a perimeter fence that visitors cannot cross until certain precautions have been taken; for example, removing footwear and wearing overalls and boots supplied by the unit. Bulk feed bins are sited next to the perimeter fence in order that they can be filled whilst the lorry stays outside the unit. Stock for sale have to walk along a loading ramp that leads outside the perimeter fence to avoid livestock lorries bringing dung onto the unit. These precautions are not usually possible for cattle and sheep units as the animals graze outside.

However, cattle units should ensure that the field fences are stock-proof to stop neighbours' cattle wandering on to the farm. This is not so easy to do for sheep as they seem to be able to spot a weakness in the best of fences.

Pests

Rats and mice not only spread disease but also cause a lot of contamination and damage. Livestock units should have a rodent control policy. Some private companies offer a contract whereby they will visit the unit on a regular basis, check for rodents and put down bait where necessary.

Whilst dogs and cats are part of farm life and, in the case of dogs, can be extremely useful to the stockperson, they must not be allowed to drag afterbirths or dead animals on to or around the unit. Feed bins and feed barrows should be covered to stop birds contaminating the food. Wire mesh can be put over the windows of intensive livestock buildings to stop birds getting in.

Prevention of disease spreading within a livestock unit

Environment

The environment in which the animal lives probably has the greatest influence on the continuing good health of the animal. Because of its importance, Chapter 13, entitled Housing and Health, deals exclusively with this topic.

Management

It is very difficult to define what 'good management' is, as the phrase encompasses every aspect of keeping and looking after livestock. Maintaining the correct environment, providing adequate nutrition and ensuring the animals are not put under any undue 'stress' can all be described as good management.

The role of the livestock 'manager' is to maintain the delicate balance between the health of the animal and the pressure of other factors that are working against the animal to cause disease, such as poor environment, other sick animals and inadequate immunity.

Hygiene

The more intensive the livestock system, the more attention must be paid to hygiene. The stockperson must avoid sloppy work practices and pay attention to detail. Simple things such as feeding sick animals last, so that infection does not spread, and scrupulous cleaning of all feeding utensils, all help to reduce the spread of disease.

The animal's own dung is a constant source of infection, so one of the animal's major requirements is a clean lying area. Many livestock buildings and pens require regular cleaning and disinfection, a topic that is dealt with in detail on page 99.

Isolation of sick animals

Sick animals should be isolated from the rest of the herd or flock and placed in a pen, paddock or yard close to where the stockperson can have regular contact. This not only removes a source of potential infection from the other animals, but also ensures that the animal is close by for treatment if necessary.

On a practical note, it is desirable to have isolation pens with a door that opens outwards as sick animals have a habit of lying against the door.

Drug therapy

The spread of and the severity of a disease can be reduced by the use of drugs. The best examples of these are antibiotics, insecticides and wormers (anthelmintics).

Antibiotics

Antibiotics kill bacteria either by damaging the cell or preventing multiplication of the bacteria. They are administered under the guidance of a veterinary surgeon to treat a specific problem or disease. The normal mode of administration is by intramuscular injection, but they can also be included in the animal's ration. This 'in-feed' medication is often used to control a disease that has affected most of the herd, as it is the most effective way of administering a drug to a number of animals. However, it is important to remember that sick animals have a poor appetite so may not receive the correct dosage of the drug.

Indiscriminate use of antibiotics over a long period, especially at low levels, can induce the bacteria to mutate, forming a resistant strain. The resistant strain will then pass on its resistance to other bacteria as it multiplies, rendering that drug useless for the treatment of a particular illness.

Meat and milk from animals treated with antibiotics must be withheld from sale until the treatment has finished and the drug has been broken down by the animal's body, leaving no residues that can be passed on to humans. It is for these reasons that antibiotics must only be used under veterinary guidance.

Insecticides

Insecticides are used to kill insects and are available as sprays, powders and the so-called 'pour-on' preparations. This latter group are applied to the skin surface, absorbed through the skin and enter the blood stream. They are carried throughout the whole of the animal's body and are effective against such things as lice and mange.

Wormers (anthelmintics)

Wormers are effective against a whole host of internal parasites such as lungworms, liver fluke and intestinal worms.

Under this section of drug therapy no mention has been made of the treatment of disease caused by virus. This is because at the moment there are no drugs available that are effective against virus diseases in animals.

Regular veterinary visits

Regular visits from the veterinary surgeon can go a long way in preventing disease. Many farmers and stockpeople look upon the veterinary surgeon as someone to be called on as a last resort, only available to treat sick animals. But regular visits, say every three or four months, when the veterinary surgeon and the stockperson can get together to look at and talk about the health of the unit, can help build up a health profile of the unit. This information can be useful when trying to identify or locate the cause of a problem.

Chapter 13

Housing and health

Since the late 1950s more and more animals have been kept in smaller areas, mainly by the use of so-called 'intensive' animal housing systems.

Most farmers and stockpeople would agree that this increase in intensification has been accompanied by an increase in disease problems. This can be explained, in part, by better reporting of disease when it occurs and more research into disease. The remainder of the increase, however, demonstrates how the environment plays a vital role in the health of the animal. As every stockperson knows, a poor environment can increase the susceptibility of an animal to disease and can aid the multiplication of disease-causing organisms.

Intensification has some positive advantages for the animal in terms of better nutrition and a reduction in the climatic effects. The obvious benefits for man are better management of the stock, better working conditions and increased production efficiency. Unfortunately, these advantages are occasionally outweighed by the disadvantage of an upsurge in levels of disease, the cost of its control and the cost of lost production.

Animal welfare is now a major consideration when planning a new building or converting an old one. The aim of this chapter, however, is to examine the relationship between animal housing and animal health and to suggest ways that a stockperson can manage a livestock house to improve health. The issue of animal welfare is dealt with in Chapter 14.

The housing requirements of the animal versus those of the stockperson

Any building designed for livestock has to satisfy the needs not only of the animals that will use the building, but also the needs of the stockperson who looks after the animals. The building designer has a difficult task as the two sets of needs can sometimes cause conflict, although good design will usually offer an acceptable compromise. Table 13.1 shows the basic housing requirements of the animal compared to those of the stockperson.

Basic types of livestock buildings

Different classes of stock require different environmental conditions. In the United Kingdom, dairy cows can thrive in the most rudimentary of buildings, with the

TABLE 13.1 *The housing requirements of the animal and the stockperson.*

Needs of animal	Needs of stockperson
Feeding area	Easy to provide feed
Lying area	Ease of cleaning/disinfecting
Dunging area	Access for machines if necessary
Lounging area	Ease of maintenance
Adequate ventilation	Holding/catching facilities
Clean water	Adequate ventilation
Easy entry and exit	Adequate light
Lighting	Ease of dung removal

temperature being of little importance. Piglets, on the other hand, need to live in a warm, draught-free environment, whilst sheep and lambs can tolerate almost anything but chilling rain. Most farm buildings are therefore designed to reflect the needs of the animal.

Animals are housed in one of three basic types of housing:

Housing that is influenced by the climate;
Controlled environment housing;
Kennel or monopitch housing.

Housing influenced by the climate

This is the most basic type of farm building, whose function is simply to provide cover and protection from the elements. It is most suited to animals that are able to adapt and thrive under a range of climatic conditions. This type of building has reasonably low stocking density and has natural ventilation. It is suitable for cattle over five to six months of age, for sheep and for adult pigs as long as generous bedding is provided. This bedding allows the pig to bury itself and create its own microclimate.

The building will normally consist of a large span-type construction with a ridge ventilator in the roof. The sides are block built to about 2 m above floor level with slatted boarding (see Fig. 13.1) or a large movable curtain forming the bulk of the side wall. They are cheap to construct (in comparison to more specialized buildings) and can often be built by the existing farm workforce.

Controlled environment housing

These provide the animal with a suitable microclimate within the building (see Fig. 13.2). They are usually fitted with some form of powered ventilation, some form of internal heating and, under certain conditions, some form of cooling of air that goes into the house. They are suitable for animals that are fed concentrated rations, so that the animal's energy output in the form of heat generation can be kept to a minimum. Stocking density is usually high, with the resulting risk of the spread of disease.

Controlled environment housing requires sophisticated management. It may be the lesser of two evils to accept slightly lower production levels in more basic types of buildings, than to run the risk of a disease outbreak in a controlled environment building.

FIG. 13.1 An open yard type of building suitable for cattle.

FIG. 13.2 A controlled environment house used for finishing pigs. The boxed sections on the roof house extractor fans for the ventilation system.

Kennel or monopitch housing

This type of housing provides the animal with a closed-in lying area that restricts the flow of air and uses the animal's own body heat to maintain the temperature (see Fig. 13.3). The remainder of the pen will be the dunging and lounging area (with provision for clean drinking water) and will have a large air space that is influenced by the climate. The building is usually divided into pens, allowing small groups of animals to be kept together. If these pens allow no physical contact between the groups and have a separate air space to the adjacent pens, the risk of a disease sweeping through the whole building is significantly reduced. The advantages of this type of building are:

It is reasonably cheap and can often be built using farm labour;
It has some of the advantages of the controlled environment housing but with less risk of disease spreading through the whole building;
The animals are able to choose the most comfortable living area, inside when the weather is cool, outside when the weather is warm.

These buildings are suitable for all classes of stock, including weaned calves, dry sows, growing pigs, beef animals and ewes and lambs.

Building management to maintain good health

A building is only as good as the management it receives. Animals in the best designed buildings can suffer from disease if some basic rules are ignored.

FIG. 13.3 A monopitch building used for weaned calves.

Stocking rate

Stocking rate is the number of animals kept in a building or pen. Buildings are designed to have space for a specific number of animals, which is calculated by allowing a number of square metres per animal. The minimum space allowances for livestock are stated in the Codes of Recommendations for the Welfare of Livestock (see Chapter 14, Animal Welfare).

Problems arise however, when animals get bigger and heavier. A pen that is just big enough for fifteen pigs with an average weight of 25 kg per pig will not be big enough three weeks later when the average weight per pig will have risen to about 35–38 kg. So it may be that the number of animals in a pen is reduced as they grow, effectively increasing the space per animal.

Overstocking can lead to many immediate problems such as fighting amongst the animals and competition at the feeding and drinking areas. However, the long-term problems can be more serious. The stress caused by overstocking can be a contributory factor to disease, which will spread quickly in an overcrowded building. Problems can be avoided by ensuring all the animals have enough space to sleep and move around and ensuring that the animals in each pen are matched for size and weight in order to reduce bullying.

Regular destocking

Regular removal of all stock from a building ensures that the disease-causing organisms have no animal host in which or on which to live.

The building is then able to be cleaned and disinfected—a topic dealt with in more detail later in this chapter.

Unfortunately, the so-called 'all-in, all-out' policy (whereby a building is filled with stock then totally emptied) can be difficult to achieve in practice, as animals are not respecters of plans and will come on heat, calve and farrow to their own timescale.

Building maintenance

At first glance it is difficult to relate building maintenance to disease on the unit. But as the stockperson is increasingly reliant on mechanical services to help with livestock management, so he or she must appreciate that any breakdown in these services can have an adverse affect on the health stock. The following list is a 'mini' check list to illustrate areas of the building that may require maintenance.

1. Doors, gates and windows should close effectively to reduce draughts and ensure that no animals can get out (or in).
2. Roofs and gutters need to be inspected regularly to avoid water dripping on the stock and soaking the lying area.
3. Floors must be rough enough to stop the animals slipping, but not so rough as to cause damage to hooves and legs. In addition, cracked and broken concrete floors can harbour disease-causing organisms.
4. If the floor has concrete slats, they should be inspected for cracking and

chipping on the edges. Worn slats can quickly cause severe foot and leg damage.

5. All feed troughs and feed fences should be cleaned regularly and old, stale food removed. They must be in good condition with no sharp edges or projections to injure the stock.
6. All electrical fittings must be checked regularly by a qualified electrician.
7. Ventilation systems require regular cleaning to remove dust. All ventilation controls such as thermostats and fail-safe mechanisms need to be checked to ensure they work effectively.
8. Automatic feeding systems that deliver the food by weight need checking to ensure accuracy.

Ventilation

Before starting this section I should say that, having had experience of a range of ventilation systems, I would advise anyone contemplating the installation of such a system, or having a problem with an existing one, to consult a specialist engineer or advisor.

The ventilation of livestock buildings is an extremely complex subject. The object of this short section is to suggest why buildings are ventilated and to describe some basic types of ventilation systems.

The functions of a ventilation system are:

1. To remove disease-causing organisms (usually respiratory) from the atmosphere.
2. To avoid a build-up of harmful gases in the atmosphere.
3. To reduce the concentration of dust in the atmosphere that can be harmful to both animals and stockpeople.
4. To remove humid air that may cause condensation within the building.
5. To help control the temperature within the building.

Most of these functions are self-explanatory, but the points regarding 'harmful gases' and 'condensation' need further explanation.

Harmful gases

When animals breathe they require oxygen and they expel carbon dioxide. Inside the building however, they also excrete dung and urine which, as it decomposes, can lead to a build-up of gases more harmful than carbon dioxide. The following is a list of gases that can be detected in livestock buildings. (The amounts of gases will vary according to type of stock, stocking rates and type of building.)

Carbon dioxide — colourless, odourless and heavier than air.

Ammonia — colourless, but with a pungent smell. It is released from dung and urine. Buildings with slatted floors and a slurry channel below have less of an ammonia problem than buildings with solid floors, as the gas dissolves readily in the water of the slurry channel. The gas is irritating to the eyes and throat of the stockperson, but in normal concentrations has little adverse affect. Very high concentrations can cause damage to livestock.

Hydrogen Sulphide — colourless, with the characteristic smell of 'rotten eggs'. It is formed by decomposing dung and can reach high concentrations when slurry pits are agitated or emptied. It is very toxic to both animals and man.

Methane — colourless, odourless and lighter than air, it is formed by decomposing dung. Since it is lighter than air, it does not usually affect livestock, but as methane is highly inflammable there is a risk of explosion from pockets of methane in slurry pits and dung channels.

Carbon monoxide — not usually found in livestock buildings. However, a building that relies on gas burners for heating, for example in piglet creep boxes, may show some signs of carbon monoxide if the burners are not maintained in good condition.

Condensation

The amount of water vapour that air will hold increases as the temperature of the air rises. The temperature of air inside a livestock building will tend to rise due to the heat generated by the animals. However, the animals not only give off heat but also water in the form of urine, sweat and water vapour in exhaled breath, so the relative humidity of the air inside the building begins to rise.

When the temperature falls, usually in the evening, the result is that condensation will form on the roof and walls inside the building, dripping down on to the stock and bedding, causing discomfort to the animals. This may force the stock to move and avoid the damp, cold areas of the building which results in uneven distribution of the stock. It is interesting to note that this combination of low temperature and high humidity, which is often the case in livestock buildings during the winter months, encourages the growth of infectious organisms, especially those that affect the respiratory system.

Condensation causes another problem, that of deterioration of the fabric and fittings of the building. Window frames rot and metal stanchions and gates soon show signs of rust. Also, it is extremely dangerous if condensation gets into electrical fittings such as light switches or fan motors as it can cause short circuits leading to fires and possible electrocution.

Types of ventilation systems

Ventilation systems can be divided into two basic types:

1 Natural ventilation.
2. Powered ventilation.

Leading on from these two types are a host of variations and refinements that are too complex to be dealt with in this book. The various advantages and disadvantages of some basic types of ventilation systems are described in the following section, with diagrams to explain the airflow patterns.

Natural ventilation

This system uses the so-called 'stack effect' to achieve its aim. When the animal's body heats the air inside the building, hot air rises to outlets in the roof and clean, cold air from outside enters the building through slatted boarding or hopper-type windows (Fig. 13.4).

The obvious advantage is that it is cheap and easy to operate. Its main disadvantage is that it is not automated and does not respond to changes in either the internal or external temperature. If the inlets and outlets are correctly sited, this type of system can produce an even temperature pattern throughout the building, so avoiding hot or cold spots which can cause discomfort to the animal.

A development of natural ventilation is a system called automatic control of natural ventilation (ACNV). This involves the use of a thermostat connected to a motor that will progressively open or close the air inlet or air outlet flaps according to the internal temperature. This overcomes one of the disadvantages of natural ventilation, that of automation, whilst still keeping a simple system.

Powered ventilation

Powered ventilation relies on fans to either extract air from a building or draw air into a building (Fig. 13.5). It is used mainly for intensively housed stock, but cattle yards with high stocking rates or ones that have restricted air flow around the building can use powered ventilation quite successfully.

The advantage of these powered systems is that they are automated and usually connected to a thermostat and a variable speed fan, so can react quickly to changes in temperature, thus maintaining a constant temperature inside the building.

The disadvantages are that running costs may be expensive and the system requires regular maintenance.

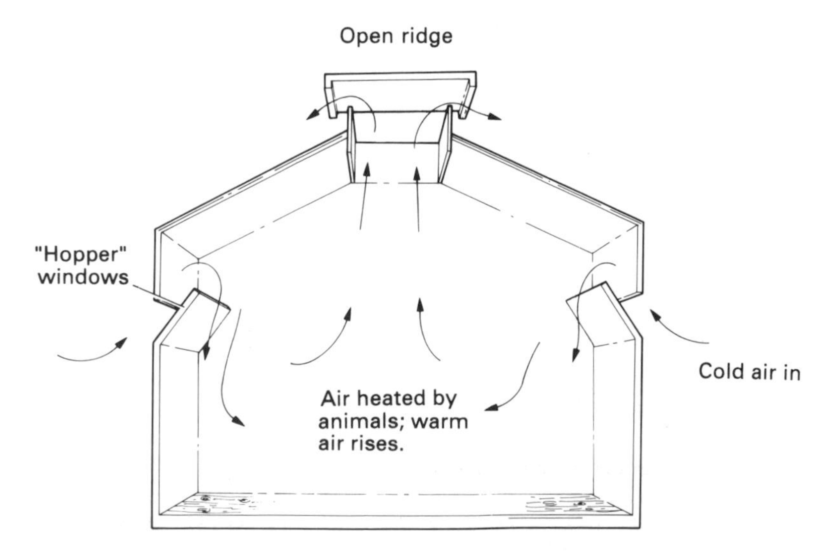

FIG. 13.4 Natural or 'stack effect' ventilation.

A refinement of powered ventilation is pressurized ventilation. For this system air is 'forced' into the building either through a duct running along the ridge or above a false ceiling (Fig. 13.6). The outlets are on the side walls. Because the inside of the building is in effect 'under pressure', it is difficult for draughts to enter. This system is especially useful for farrowing houses to avoid cold draughts on baby piglets.

It should be noted that any livestock building that relies on powered ventilation to supply air to the animals must have an alarm to warn the stockperson of a system failure and some non-mechanical means of ventilation. This avoids the possibility of livestock suffocating in the event of a power cut or breakdown.

Cleaning and disinfection of livestock houses

An efficient method of disease control is to improve standards of hygiene rather than to rely on drugs. Part of the overall hygiene strategy of a livestock unit should be the regular cleaning and disinfection of all livestock buildings. If done properly, this

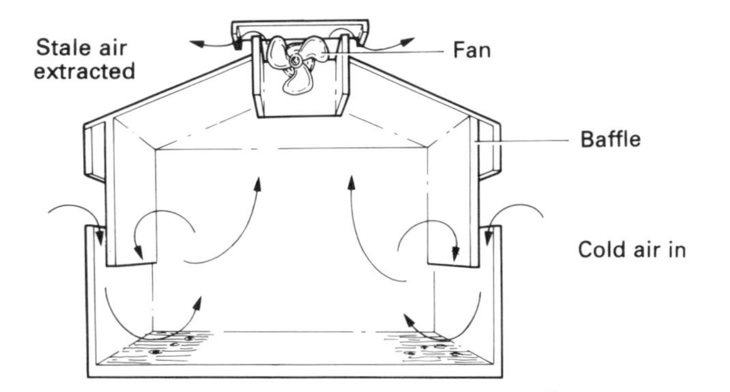

(a) Air extracted from building

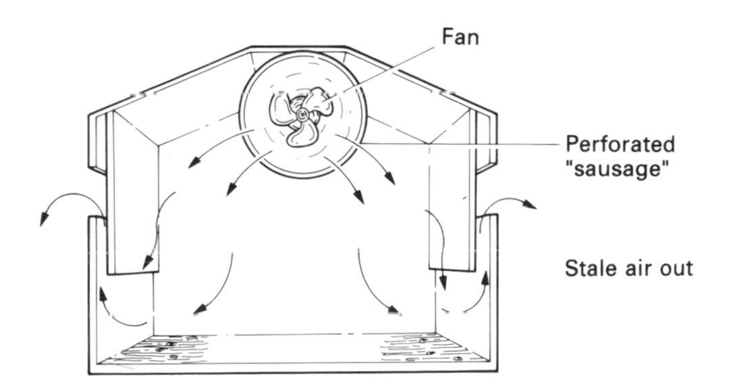

(b) Air drawn into building

FIG. 13.5 Powered ventilation. (a) Air extracted from building. (b) Air drawn into building. The perforated 'sausage' runs the length of the roof. An exterior opening draws in air from outside and a fan delivers it the length of the 'sausage'.

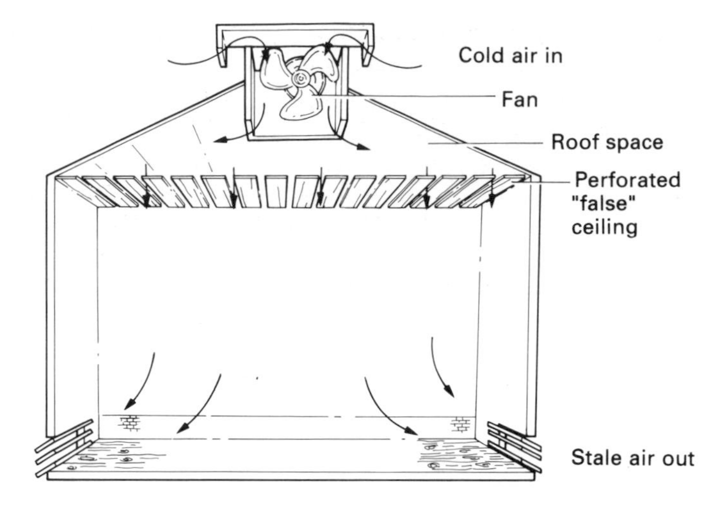

FIG. 13.6 Pressurized ventilation.

should remove all disease-causing organisms and render the building free from infection.

A disinfectant is an agent that destroys micro-organisms. This could be the heat from a flame gun, a steam cleaner, sunlight, cold or a chemical. Livestock farming usually relies on chemical agents, although steam cleaners, flame guns and, to a certain extent, sunlight are still used.

There are chemical disinfectants that kill bacteria—bactericides, that kill viruses—viricides, and that kill fungi—fungicides. Some chemical disinfectants will kill most forms of micro-organisms whilst others are more selective. There is none that is one hundred per cent effective. The label on the disinfectant container will usually say what the product should be used for and whether it is effective against bacteria, viruses or fungi.

The Diseases of Animals (Approved Disinfectants) (Amendment) Order, 1988, lists a large number of disinfectants and the concentration that they should be used at in order to be effective at killing some specific micro-organisms; for example, the foot and mouth virus.

Disinfection check list

The following is a check list that should be followed when a building requires routine disinfection. It assumes that no contagious or infectious disease has been present. If a serious outbreak of disease has occurred, then cleaning, disinfection and disposal of the dung and bedding should be carried out under veterinary guidance.

1. Remove all fittings and equipment such as feed hoppers, feed racks and gates. These should then be cleaned of all organic matter and soaked in disinfectant.
2. Remove all dung and bedding from the building.
3. The roof or ceiling, beams, pillars and ledges should be cleaned of dust using an industrial vacuum cleaner.

4. Isolate the buildings from the main power supply to avoid the danger of electrocution, as water could accidentally contaminate the electrical fittings.
5. Thoroughly clean the lower part of the walls, the floor and any immovable fittings with a detergent disinfectant. The most efficient method of doing this is to use a pressure washer. (Note: As the electricity supply to the building should have been cut off (see 4), the machine will usually need an extension cable. Great care must be taken not to get water on to this cable as there is a real risk to life if water enters the electrical socket. A circuit breaker or low voltage power supply should be used for extension cables).
6. If the building has earth floors, they may be soaked in a disinfectant.
7. After the building has been cleaned and all organic matter washed away, soak the building with an approved disinfectant at the correct strength.

Fumigation

Fumigation of the building can be carried out in addition to or in place of the final disinfection. Fumigation is a popular method of disinfection as it is cheap and effective. The gases used are usually toxic to humans, so the appropriate safety precautions have to be taken, e.g. the use of respirators.

Fumigation can be achieved by the use of 'smoke'—that is mixing two chemicals whose chemical reaction is to produce a smoke, or by fogging—that is using a machine to produce a fine mist of the chemical solution that will cover the interior surfaces.

Fumigation is a specialist procedure, so appropriate advice should be sought before undertaking any disinfection using this method.

Outdoor systems

In general terms it is true to say that outdoor (extensive) systems of keeping animals have fewer health problems than indoor (intensive) systems. That is not to say that outdoor systems never have severe problems. They certainly do, but on the whole they are relatively healthier.

In the United Kingdom most dairy herds and sheep flocks, although intensively managed for maximum production, are housed under extensive conditions. Apart from the winter months, cows live out in the fields and most sheep live outside for the majority of the year although some are housed, especially at lambing time.

Pigs, however, are normally kept under intensive conditions, with the minority being kept outside. Although this section could deal with all classes of livestock, it will concentrate only on pigs, as this seems to be the area where most progress on outdoor systems is taking place.

Pigs outdoors

The past five to eight years have seen a lot of interest being shown in outdoor systems (see Fig. 13.7). It is interesting to note the factors that have brought about this revival.

Costs

Building costs for new pig housing have risen dramatically in the past few years. Prices for a new unit are quoted as high as £1600 per sow place (1990), with interest charges being extremely high. Building costs for an outdoor unit are quoted as £160 per sow place, but this does not include any cost for the land involved. However, in the light of falling land prices and restrictions on production of milk and possibly cereals, land used for pigs could be an attractive alternative. Electricity and gas costs have all risen in relation to feed costs, as have repair and maintenance costs.

Green issues

Increasing concern over the countryside in general and in particular the way in which farm animals are kept has led to a re-examination of outdoor systems. Large-scale pig farms have some difficult issues to face up to including:

Animal welfare.
Storage and disposal of pig slurry or solid muck, especially close to urban areas.
Planning regulations that may limit possible expansion.

FIG. 13.7 Dry sows managed under an extensive outdoor system.

Outdoor systems are perceived as 'animal friendly', have few slurry or muck disposal problems and, as little building work is involved, are not subject to such rigorous planning regulations as their more intensive cousins.

Advances in the pig industry

Advances in technology, changes in consumer demand and, not least, economic pressures have forced a number of changes on the pig industry. In their own way these changes have made outdoor pigkeeping a more viable proposition. These changes include:

1. Computer-controlled sow feeding, which allows for accurate, automated feeding. This is perhaps not so applicable in an outdoor situation, but certainly suitable for loose housed sows. Computer recording programs for outdoor herds have led to more efficient management of individual sows and of the unit as a whole.
2. Improvements in fencing, especially electric fencing, means that fields can quickly be split into smaller paddocks, enabling matched groups of sows or gilts to be kept together.
3. Water, the bane of outdoor pigkeeping, can now be delivered to a field or paddock very efficiently because of advances in polythene pipe jointing. This allows pipes to be joined by hand so the tool box, full of giant spanners and wrenches, can be left in the workshop.
4. Pig feed manufacturers now produce feed in roll or biscuit form, making it easier for the stockperson to feed with less waste.
5. Pig breeding companies will supply gilts that have been specially bred for outdoor systems. The breeding background to these gilts is usually based on the British Saddleback, crossed with either a Large White or Landrace.

Health on the outdoor system

Turning pigs out into a field is no panacea for health problems. The farmer may simply be substituting one set of troubles for another. Strict precautions need to be observed as on any livestock unit. These include keeping new stock in quarantine for a period of time, isolation of sick animals and vaccination programmes against any known disease. (See Chapter 12, Disease Prevention.)

Whilst there are usually fewer environmental disease problems on an outdoor unit, there are some drawbacks that have an effect on health.

Ground conditions

Light, free draining soil is the most suitable for outdoor pigs. Clay soils lead to muddy gateways and access roads. If the soil is very stony, it may give rise to some foot and leg problems. Some outdoor systems regularly run the breeding stock through a foot bath to help avoid any bacterial infections.

Sunburn/heat stroke

Pigs have a very poor heat regulation system (*see* page 31). They do not sweat except by panting and have little or no body hair. The combination of these factors leads to a very real risk of severe sunburn or possible heat-stroke under sunny, humid conditions. Pigs outdoors must be provided with some form of shade, as their huts get extremely hot inside. Some form of muddy water hole as a wallow is essential in the summer. Pigs in a wallow will cake their skins with a layer of mud, making a very effective sun block that also cools the body by evaporation of the water.

Infertility

There is some evidence to suggest that shortening day lengths in August and September may lead to temporarily reduced oestrus activity in sows. Under intensive conditions this may be counteracted by the provision of artificial light for the same time period each day. This is obviously not possible under the outdoor system.

Periods of high daytime temperature during the summer can affect the libido of the boar and the quality of his semen, leading to an increase in the number of returns and a reduction in the number of piglets born alive. The provision of shade and wallows can help reduce the problem.

Observation

Under intensive conditions it is quite easy to inspect the stock at any time, day or night. Stock inspection on outdoor units is more difficult and takes longer, but is an essential element of the disease-prevention strategy. Feeding time provides the best opportunity for the inspection of stock, especially as one of the first signs of illness in pigs is the animal not eating.

Intensive livestock buildings are here to stay. Advances in building materials and new technology will improve the living conditions of animals and the working conditions of the stockperson. However, the first point of contact for the management of the animal will still be the stockperson.

One of the main roles of the stockperson now and in the future is to be aware of the influence that the environment has on the health of the animal. It is up to him or her to manage the building with the health, comfort and welfare of the animal as the major consideration.

Chapter 14

Animal welfare

The object of this chapter is to give some background as to why animal welfare has, in recent years, become a prominent issue both inside and outside the industry. I also hope to raise the reader's awareness of some of the arguments on both sides of the debate.

This section does not set out to define what animal welfare is, or even how it can be achieved, because results of trials and research work are all the time adding to our knowledge of animal behaviour.

To say that the agricultural industry in the United Kingdom is undergoing rapid change would be a case of classic British understatement. The past few years have seen farmers placed under massive pressure from a number of directions, including international, financial and, not least of all, the concern of the general public. These pressures have resulted in such changes as milk quotas and cereal levies and changes in the way that some farming practices are carried out.

It seems that farmers, as victims of their own success, are perceived by non-farmers (and the 'media' in particular) as spoilers of the countryside, polluters of rivers and exploiters of farm animals. 'Green' issues such as conservation and animal welfare are now seen as more important than what was the traditional role of farmers, that of providers of food. In the early 1940s getting food and not going hungry was more important than where it came from. In the 1990s, with our full stomachs and centrally heated houses, the emphasis has changed.

At the moment the British position on animal welfare is that the Ministry of Agriculture (using advice and information provided by the Farm Animal Welfare Council) has drawn up a number of Codes of Recommendations for the Welfare of Livestock, covering most classes of stock, that seek to define the basic requirements for good welfare. These are voluntary codes and contravention of them is not an offence in itself. However, in a legal prosecution, they could be used in evidence to help prove an offence.

Allied to this there are more specific pieces of legislation regarding farm animal welfare. They are those that refer to surgical operations allowed to be carried out on farm animals. The main elements of these are contained in a number of Acts of Parliament:

1. The Agriculture (Miscellaneous Provisions) Act, 1968 includes various regulations referring to:

—Intensive units (1978).
—Prohibited operations (1982).
—Use of anaesthetics (1974).
2. The Protection of Animals (Anaesthetics) Act, amended 1982.
3. The Veterinary Surgeons Act, 1966 with a number of exemptions and amendments (1982).

Tables 14.1, 14.2 and 14.3 set out the current legislation relating to some of the more common on-farm surgical operations, including age of animal, the methods allowed, who may perform the operation and whether anaesthetic is required or not.

In the future it looks as though animal welfare will see a shift in emphasis from voluntary codes to more direct legislation. Already there are The Welfare of Calves Regulations, 1987, designed to outlaw restricted individual crates for veal calves, and Holland is committed to banning battery cages for laying hens from 1994, providing suitable alternatives have been developed by then.

It should be pointed out that any restrictions in one country may put their farmers at a disadvantage to their trading partners, so any legislation that requires radical changes in farming practices could lead to some sort of common European approach.

In response to this possibility, advisory bodies, companies and farmers have put a lot of effort into the research and development of new farming techniques. An example of this in the pig industry is the development of the computer-controlled electronic sow feeder. This allows pregnant sows to be group-housed (see Fig. 14.1), often on deep straw, whilst still maintaining the accuracy of feeding that is so vital during pregnancy. For the welfarist it is certainly a more attractive alternative to the confinement of sows in stalls and tethers, but the system is not without its problems. Bullying amongst the sows and difficulties in spotting sick animals amongst such large groups are just two such examples.

FIG. 14.1 Pregnant sows, group housed on deep straw. Accuracy of feed levels are maintained by use of computer-controlled feeding stations that recognize individual sows by means of their electronic ear tags.

TABLE 14.1 *A summary of the law relating to common on-farm surgical operations allowed to be performed on cattle (as at March 1990).*

Operation	Age of animal	Methods	Person who may perform	Anaesthetic required Yes/No
Castration	1. First week of life *only*	Rubber ring or other device to constrict the flow of blood to the scrotum	Trained stockperson	No
	2. Up to 2 months	Other than above	Trained stockperson	No
	3. 2 months and over		Veterinary surgeon	Yes
De-horning	Any age	Not specified	Trained stockperson	Yes
Disbudding calves	1. First week of life only	Chemical cauterization	Trained stockperson	No
	2. Unspecified	Other than above	Trained stockperson	Yes
Removal of supernumerary teats of a calf	1. Up to 3 months	Not specified	Trained stockperson	No
	2. 3 months and over		Veterinary surgeon	

(Source: ADAS).

TABLE 14.2 *A summary of the law relating to common on-farm surgical operations allowed to be performed on pigs (as at March 1990).*

Operation	Age of animal	Methods	Person who may perform	Anaesthetic required Yes/No
Castration	1. First week of life *only*	Rubber ring or other device to constrict the flow of blood to the scrotum	Trained stockperson	No
	2. Up to 2 months	Other than above	Trained stockperson	No
	3. 2 months and over		Veterinary surgeon	Yes
Docking of tails	1. First week of life *only*	Quick and complete severance of the part of the tail to be removed	Trained stockperson	No
	2. Unspecified	When in the opinion of a veterinary surgeon, the operation is necessary for reasons of health or to prevent injury from the vice of tail-biting	Veterinary surgeon	Yes

(Source: ADAS)

TABLE 14.3 *A summary of the law relating to common on-farm surgical operations allowed to be performed on sheep (as at March 1990).*

Operation	Age of animal	Methods	Person who may perform	Anaesthetic required Yes/No
Castration	1. First week of life *only*	Rubber ring or other device to constrict the flow of blood to the scrotum	Trained stockperson	No
	2. Up to 3 months	Other than above	Trained stockperson	No
	3. 3 months and over		Veterinary surgeon	Yes
Docking of tails	1. Any age	PROHIBITED unless sufficient tail is retained to cover the vulva of female sheep and anus of male sheep		
	2. First week of life *only*	Rubber ring or other device to constrict the flow of blood to the tail	Trained stockperson	No

(Source: ADAS)

What is 'animal welfare'?

The phrase 'animal welfare' has, in the past few years, taken on several different meanings. To some it means that animals should not be slaughtered for meat, to others it means animals should be reared in fields; others define it as allowing animals to express their natural instincts. Animal welfare encompasses far wider issues than these. The health of the animal is central to welfare, with the efforts of the stockperson being directed towards the maintenance of good health. The simplistic concept that free range is good and intensification or 'factory-farming' is bad takes no account of the stress animals can suffer from, for example, severe weather and serves only to over-simplify a complex subject.

A good starting point when trying to define animal welfare is to look at the Codes of Recommendations for the Welfare of Livestock. These state good welfare as: '. . . a husbandry system appropriate to the health and, so far as is practicable, the behavioural needs of the animals and a high standard of stockmanship'.

The prefaces to the Codes also detail recommendations that, if followed, help to fulfil the physiological and behavioural needs of animals. These include:

1. Providing physical comfort (possibly bedding?), shelter, fresh water and an adequate diet.
2. Providing light during daylight hours and having adequate lighting to enable the stockperson to inspect the animals at any time.
3. Freedom of movement and adequate space for that movement.
4. Having access to other animals, particularly of the same species.
5. The prevention and treatment of disease or injury.
6. Providing some kind of emergency arrangements to cover the event of fire or disruption of services such as electricity and breakdown of essential mechanical services; for example, ventilation systems.
7. Avoiding unnecessary surgical procedures such as castrating and tail-docking.
8. Providing the animal with the opportunity to exercise most normal instinctive patterns of behaviour.

The above points can be viewed as the general guidelines, with the Codes of Recommendations being the details of how these guidelines can be implemented on the farm. All farms should have a copy of the appropriate Code available so that the stockperson (employer or employee) is aware of its recommendations.

The role of the stockperson

The stockperson is the first point of contact between the animal and its well-being. He or she has a responsibility to be able to interpret the signs that are associated with the health of an animal or a group of animals. These signs can be divided into four main categories:

1. The physical appearance of the animal.
2. The behaviour of one or more animals in relation to the group.
3. The productivity of the animal(s) (e.g. milk yield, weight gain).
4. The state of the environment.

The stockperson will monitor these signs (often subconsciously) as he or she follows the normal work routine. Any deviation from normal will alert him or her to take a closer (conscious) look at the animal(s).

The stockperson has then to decide if the animal is healthy or unhealthy. If the animal is unhealthy he/she must decide what action to take and whether a veterinary surgeon needs to be called. In order to be able to take these decisions, often involving many hundreds of pounds' worth of animal, a stockperson must have a knowledge of the common diseases and, if the disease is treatable by the stockperson, have the necessary skills to be able to perform the treatment. This knowledge comes from proper training and a wide experience of farm livestock.

Large livestock enterprises employing many people must have clear lines of communication with the various responsibilities of how to cope with an emergency well defined. One of the common criticisms of large livestock units is that too many animals are looked after by too few people. In this type of situation it is essential that the stockperson has the correct facilities and the support of the unit manager or owner in order to carry out his or her duties effectively.

Conclusion

I am aware that this chapter raises many questions but answers none. Because this subject is clouded by prejudice, common practice and long held customs, there are no clear-cut answers.

However, of one thing I am certain, the animal welfare debate will not, and should not, go away. The issues mentioned briefly here will play a large part in the agricultural industry of the future.

Chapter 15

Lameness in cattle

All classes of farm livestock suffer from some form of lameness at one time or another. However, this chapter is devoted exclusively to cattle, as lameness amongst cattle, especially dairy cows, can lead to big financial losses on some farms. Farmers who have monitored foot problems in their dairy herds have shown that 60 per cent of cows suffer from some form of foot trouble. In a hundred-cow herd it is estimated that the possible cost in lost production and treatment could be near £3000 per year. Allied to this, many people are becoming concerned that lame cows are a 'hidden' welfare problem.

Diseases of the foot

The various areas of the foot that are referred to in the following text are shown in Fig. 15.1. It will be useful also to refer to Fig. 1.5, Section through hoof of a cow, page 8.

Foul in the foot

This disease affects the soft tissue between the two claws of the hoof. It is caused by the bacterium *Fusiformis necrophorus* entering the tissue via a cut or abrasion. The animal is usually quite lame and looks in pain. Close inspection will reveal an offensive smell. As it is a bacterial disease, antibiotics administered under the supervision of a veterinary surgeon will normally clear up the infection.

Prevention:

Regular foot baths containing a cleansing agent, e.g. copper sulphate, zinc sulphate or formalin.
Ensure slurry is regularly removed from passages.
Keep concrete tracks in good repair.

Sole ulcer

These are abscesses deep in the foot, usually occurring near the junction of the sole and heel. The horn around the ulcer is weak and as the ulcer reaches the horn on the

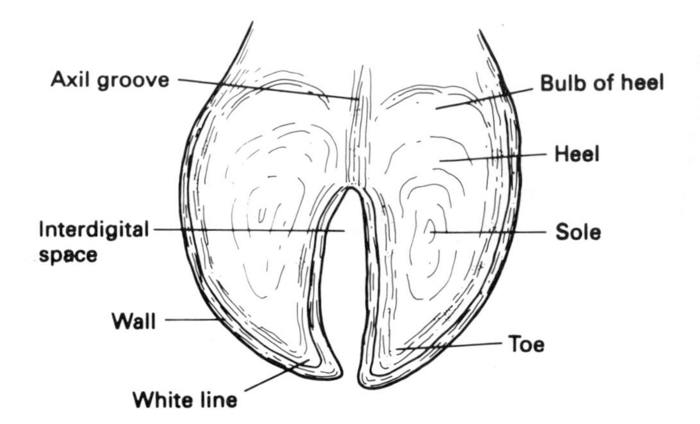

FIG. 15.1 The underside of a cow's foot.

sole of the foot it quickly breaks down, causing lameness. When the horn is pared away, a lump of flesh about the size of a redcurrant can be seen. Sole ulcers are probably caused by pressure on the foot owing to standing on hard surfaces. They are more common among housed cattle. Heifers are very prone to sole ulcers caused by the many changes they face when introduced to the adult herd, e.g. concrete surfaces, cubicles and changes in feeding.

Prevention:

Regular preventive foot trimming and foot inspection.
Ensure that cows have comfortable lying areas.

White line disease

The white line is the area on the hoof where the sole horn meets the wall horn. If this junction is weakened by laminitis or other foot problems, damage could result from cows standing on rough surfaces. Bacteria will multiply in any damaged areas and eventually pus will form causing pain and lameness. If the condition goes unnoticed, the pus will follow the white line around the hoof, sometimes covering the whole area under the sole.

Prevention:

Make sure the concrete yards and pathways are kept in good repair.
Breed from bulls with good hoof conformation.

Laminitis

The laminae is the area of the foot from where the horn grows (Fig. 15.2). It is spongy tissue with a huge number of blood vessels. If a cow's foot is trimmed too severely, one may cut into the 'quick' or laminae which will bleed profusely. Any inflammation of this laminae is known as laminitis and results in irregular growth of the horn. Initially the cow will be tender on its feet with no obvious signs of disease.

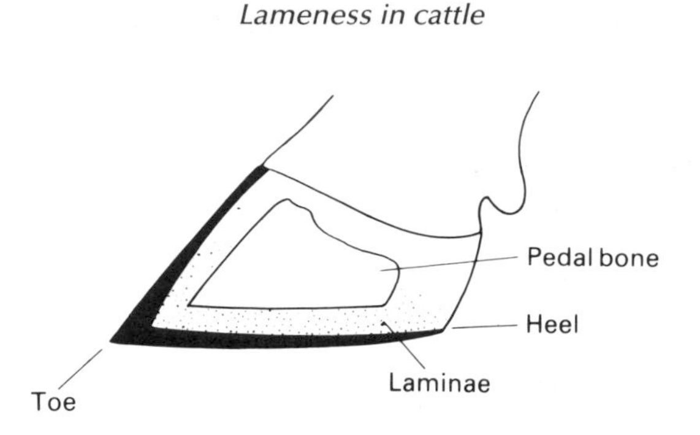

FIG. 15.2 Cross-section through cow's foot, showing the position of the laminae.

Eventually the outer wall of the hoof will form ridges or lumps, showing that the growth of the horn was abnormal. Laminitis may not be caused by one factor alone; usually a combination of conditions contribute to the inflammation of the laminae.

Prevention:

A diet that does not contain enough long fibre can lead to acidosis (*see* page 155) in the rumen, which in turn leads to nutritionally induced laminitis. Thus, modern precision-chopped silage may lead to laminitis.
Providing a comfortable lying area is essential in order to keep cows off their feet for as long as possible. Cubicles must be of the right dimensions and have some form of cushioning, for example chopped straw or rubber mat.

Heel erosion (underrun heel)

Horn on the heel is much softer than wall or side horn. It covers a pad of fibrous tissue that helps absorb shocks as the cow walks. The dirty conditions of slurry and silage effluent that the foot has to walk in means that this soft tissue is prone to bacterial attack and to the corrosive action of the acids in slurry and silage.

The heel will gradually get eroded away, which does not in itself cause lameness. However, the erosion leaves the foot open to sole ulcers and white line disease.

Prevention:

Remove slurry from passages regularly (at least twice a day).
Footbath cows in a suitable cleansing/disinfecting agent.

Digital dermatitis

This is a bacterial infection that usually affects the soft skin on the groove at the back of the foot where the two claws join. The skin is seen to be scabby and sometimes weeping, with the hairs standing erect. It is highly infectious, being transmitted in the slurry and by unhygienic foot trimming practices such as using a brush and water to clean the foot to be trimmed. It can be treated effectively by drying and spraying with 'purple' antibiotic for three days.

Prevention:

Treat infected animals promptly.
Inspect any bought-in stock for signs of the disease.

(*Note:* Digital dermatitis can affect humans. Hands should be carefully washed after handling cows' feet.)

Reducing the incidence of lameness

The preceding section outlines some ways of preventing or reducing the incidence of the particular diseases mentioned. However, in order to reduce lameness in the herd the farm must produce a strategy that does not rely on the 'fire brigade' tactics of treating only when emergencies arise. The herdsperson, the cubicle designer, the builder, the people who lay the concrete tracks, the feed supplier and the person who walks the cows in for milking are all responsible, in varying degrees, for ensuring that the cows' feet remain in good condition.

The following points need consideration when setting out to reduce lameness in the herd.

Cow comfort

Reducing the time that cows spend standing is probably the most important factor in reducing lameness. The cubicles must be comfortable (see Fig. 15.3) or cows will either not use them or will spend little time in them. There are many types of cubicle design, with much research work having gone into the various dimensions. As the genetic influence of the larger Holstein type shows in a herd, cubicles that were big enough fifteen years ago might be too small by today's standards. The average body size of the herd is worth checking if new cubicles are to be installed. Also the walking or 'lounging' area must be big enough for cows to move around in. Standing still for long periods will put undue pressure on the laminae.

Footbaths

The effectiveness of a footbath depends on how dirty the cow's foot is when it walks into the bath. If the cow's foot is caked with muck, any agent used to kill the bacteria or harden the hoof (usually formalin or copper sulphate) will not have a chance to penetrate to the site of any infection. It is better to have an arrangement whereby the bath is split into two sections: the first section containing clean water to wash the foot; the second containing the cleansing agent. Some footbaths have a short raised area between the two sections (Fig. 15.4), which allows the foot to drain off water before moving into the cleansing agent.

Slurry removal

Slurry is a relatively new invention. Until the early 1960s, cows were either loose-housed on deep straw or tied in shippons, with muck and urine being collected in a

FIG. 15.3 In order to accommodate the large body frame of the modern dairy cow, the original cubicles have been removed and the new ones installed at an angle to the dung passage; they are both longer and wider.

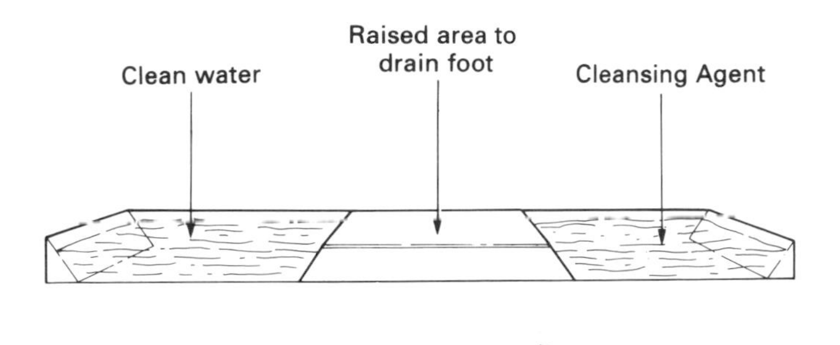

FIG. 15.4 A two-section foot bath.

rear dung channel. When cubicles became popular, cows' feet had to stand in wet cold slurry for hours on end. In conjunction with cubicles, the popularity of silage increased with cows feeding at self-feed silage faces and so standing in acidic silage effluent. In the light of the conditions that cows' feet have to function in, it is no wonder that there are problems.

Regular removal of slurry is essential to allow the horn time to dry and harden. Automatic scrapers are useful as they can be set to scrape out several times a day. Some people have observed that the build-up of slurry in front of the automatic scraper blade will cover the foot of the cow up to the coronet as it steps over the advancing blade. This can lead to hard caked muck covering the foot and so their effectiveness in keeping cows' feet clean is questionable.

Walking surfaces

All walking surfaces such as yards, passageways, farm tracks and gateways must be kept in good condition in order to reduce the effect of sharp stones on the cows' feet. When laying concrete, a round stone aggregate should be used in preference to a crushed stone one. New concrete has a corrosive action on cows' feet, so hosing the surface down or covering it with a layer of straw before the cows are brought on to it is essential.

The concrete floors of silage clamps can become very rough as the effluent erodes the concrete. A lot of farmers have used asphalt to repair these floors as it is not as abrasive as concrete, it is comfortable for the cows to walk on and it is not as slippery as worn concrete.

Nutrition

Rumen acidosis is one of the main causes of laminitis. If the diet contains little roughage, there is not much rumination of the feed which in turn leads to little production of saliva. Saliva, which is alkaline, is necessary to neutralize the acids caused by fermentation in the rumen.

The farm should adopt a feeding policy that ensures adequate amounts of carbohydrate and roughage and avoids sudden changes in the ration, such as changing from silage to spring grass which is rich in protein.

Genetics

Breeding for good feet is a long-term policy. If the herd does have foot problems caused by hoof conformation, choosing a bull to sire the herd's replacement heifers will help to correct the problem.

Regular preventive foot trimming

See following section.

Foot trimming

This section is a guide to the theory of foot trimming only and is no substitute for specialist training and practice. I strongly advise anyone who wishes to undertake the trimming of cows' feet to attend a specialist course where safe procedures and correct techniques will be taught under real conditions.

Foot trimming is an important weapon in the fight against lameness and as such it deserves a section of its own. However, overtrimming is as bad as not trimming at all. Cows' feet should be inspected at least twice a year and trimmed if necessary. Any cow showing discomfort when walking should have its feet inspected at the earliest possible opportunity.

Cows' feet grow more quickly than they get worn away, so some form of trimming is essential in order to maintain a satisfactory shape to the foot. Since feet are used for walking on they must be able to take the weight of the animal, so a look at the mechanics of how the weight is loaded on the hind legs is useful in order to appreciate the stresses that are imposed on the four claws.

Loading of hind legs and claws

A cow standing normally will have that part of the body weight carried by her hind legs distributed evenly over all four hind claws (Fig. 15.5). However, if the cow shifts its weight slightly to one side the load is spread unevenly (Fig. 15.6), the greater load being taken by the outer claw of the foot. Thus, the outer claw is more 'stressed' than the inner, leading to irritation of the laminae and excessive growth of the horn of that claw. The larger the claw, the more body load is taken on that claw, leading to more irritation and so excessive growth—a vicious circle has started. Indeed, it is usual to find an enlarged outer hind claw which needs more trimming and shaping than does the inner one.

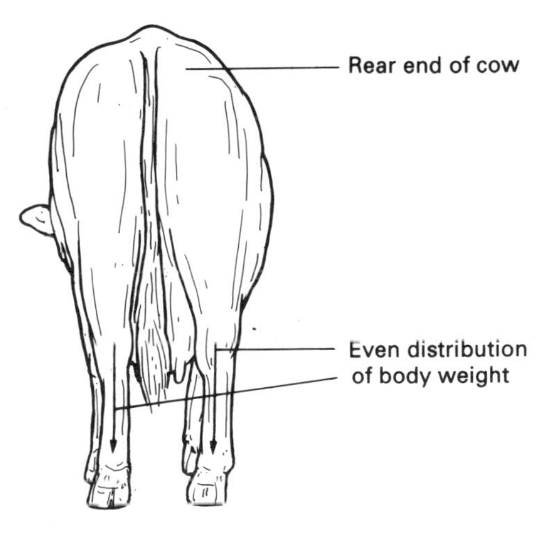

FIG. 15.5 Even loading of hind legs.

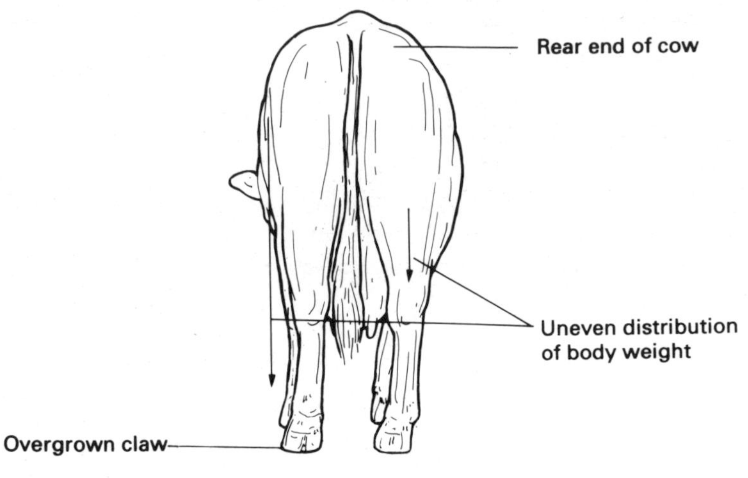

FIG. 15.6 Uneven loading of hind legs.

The aim of the foot trimmer is to trim the feet in order to give the cow flat load-bearing surfaces on which to walk, enabling the cow to spread her weight evenly over all four claws (Fig. 15.7). The trimming will also alter her posture, making her walk better and relieving any stress on her legs and hips.

Trimming

In order to trim cows' feet successfully you need:

Good cattle-handling facilities.
Sharp knives.
Plenty of practice.

TOE TO HEEL

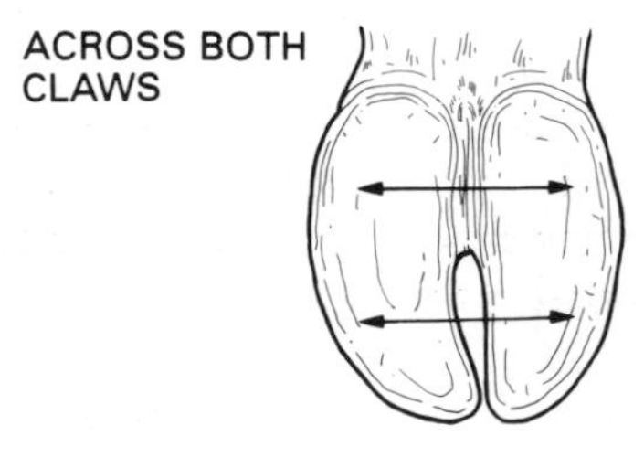

FIG. 15.7 Even load-bearing surfaces of both claws.

FIG. 15.8 Foot trimming in a specialist crush. A strap supports the belly of the cow whilst a pulley system enables the hind legs to be lifted quite easily.

The facilities should include some form of cattle crush that enables the operator to lift the foot off the ground by means of ropes or a purpose-built pulley system. There are a number of specialist foot-trimming crushes available (see Fig. 15.8) that have pullies, quick-release devices, belly straps for extra support and leg rest blocks for trimming front feet if necessary.

Method

1. Raise the foot clear of the ground until it is in a comfortable position for both the cow and the foot trimmer.
2. Clean the foot of all muck using dry wood shavings. Do not use a brush and water as this may spread digital dermatitis (*see* page 113).
3. Inspect the foot for any diseases or problems. Remove any impacted stones. Decide what needs doing to that particular foot, as all feet are different.
4. For adult Friesian/Holstein cows measure 7.5 cm from coronet to toe (Fig. 15.9). Cut off any excess growth below this point—this gives the initial shape to the foot.
5. Shave the bottom of the foot with the foot-trimming knife until the foot returns to the correct shape. A constant check must be made to ensure the depth of solar

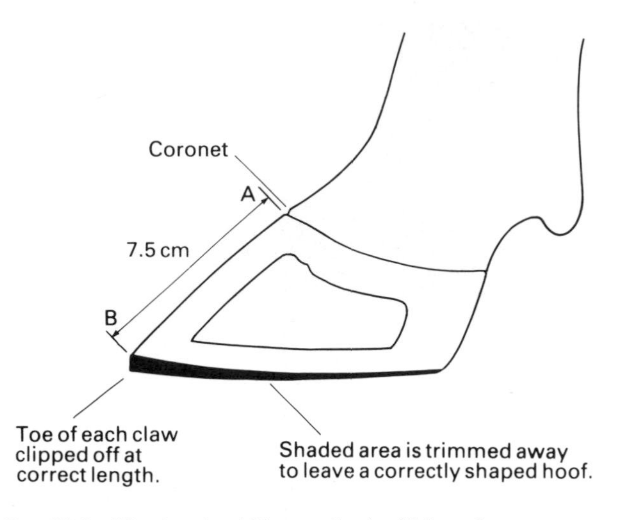

FIG. 15.9 Clipping should leave a depth of 7.5 cm from coronet to toe.

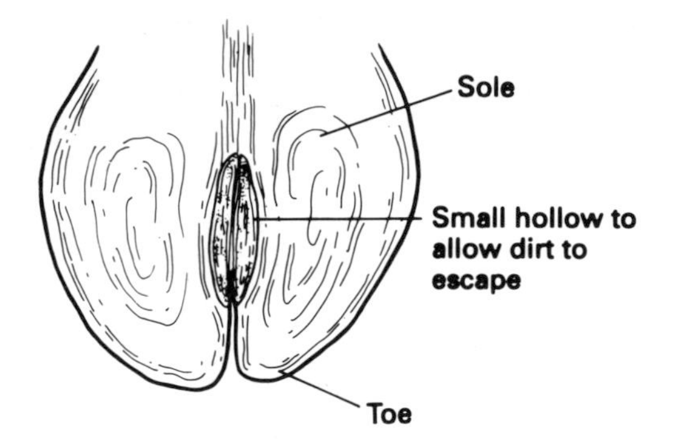

FIG. 15.10 Shaped hollow on each claw.

horn is adequate. This is done by pressing hard with the thumb on the sole: at the first sign of sponginess—stop!

6. Shape a small hollow on the inside of each claw (Fig. 15.10) to allow dirt and muck to escape.
7. Check both claws are even and that the load-bearing surfaces are flat.

If the feet are badly overgrown it may be necessary to reshape the feet over two or three trimming sessions, as overparing may cause acute lameness, especially if the soles are soft.

Chapter 16

Mastitis

The term mastitis means inflammation of the udder (or more correctly the mammary gland) and is a general term which covers most diseases and conditions of the udder. The inflammation arises as a result of either injury or infection.

The udder is prone to infection for a number of reasons:

It has an opening to the outside via the teat, allowing disease-causing organisms to enter.
Infectious organisms will thrive in a milk medium held at blood-heat.
The udder of farm animals is sited in such a position that it can easily be contaminated by faeces and dirt.

Mastitis in the dairy cow

The economic losses caused by mastitis in dairy cows are well documented and will not be discussed here. What is important for our purposes is to consider the ways in which cows contract mastitis and then to look at some preventive measures.

The severity of the disease is classified according to the symptoms it shows:

Acute—the cow will show signs of severe illness such as breathing fast and not chewing its cud. One or more quarters will be hard and hot and may have a bluish tinge.
Sub-acute—shows as small clots in the milk. This quickly progresses to show symptoms of pain and tenderness in the udder and may become acute if not treated. The milk turns yellowish and the affected quarter may dry up completely.
Chronic—there is no pain or swelling of the udder and the milk looks normal. However, there is a gradual decrease in the milk yield of the affected quarter(s).

Cows can pick up the organisms that cause mastitis in one of two ways:

Organisms transmitted from another cow.
Organisms transmitted from the environment.

In order to consider ways of prevention, it is necessary to look in detail at these two methods of transmission.

Mastitis transmitted from other cows

The main causal organisms involved in this mode of transmission are three types of bacteria:

> *Streptococcus dysgalactiae*—found in the udder and in cuts and scabs on the teat.
> *Staphylococcus aureus*—found in the udder of chronically infected cows and in cuts on the teat.
> *Streptococcus agalactiae*—found in the udder.

By far the most common mode of spread from cow to cow is associated with milking and the milking equipment, so this must be dealt with in some detail.

Milk let-down

This process is described on page 64 and involves the release of milk from the udder tissue which fills the udder spaces and the teat cistern. It is essential that milk is actually present in the teat at the start of milking, as the unit is put on to the cow. If no milk is present, the teat tissue will take the full force of the vacuum, resulting in damage to the teat.

Personal hygiene

Cracks in the milker's hands may contain bacteria that could be passed on to the cows and so rubber gloves should be worn, as these are less prone to retaining bacteria. They can be dipped in antiseptic solution between cows to reduce the risk of infection still further.

Udder/teat washing

There is much controversy regarding the washing of the udder and teats. If udders are washed, the whole area of udder and teats must be thoroughly dried, using disposable paper towels. A drop of dirty water left on the end of the teat may be a source of infection that could enter the teat during milking.

Many farmers have adopted the 'dry wipe' system of cleaning the teats, so reducing the risk of infection. However, there is an increased chance of dirty sediment contaminating the milk which may result in contravening the Milk and Dairies Regulations. If this method is used, it is important that extra attention is paid to cow cleanliness.

Foremilk check

Mastitis clots in the milk can best be spotted by stripping the foremilk into a strip cup. This is a very time-consuming job, but it does work. However, this method does run the risk of passing mastitis from cow to cow as does another common practice—that of stripping the foremilk onto the floor of the parlour.

Filter screens can be used to detect clots. These fit into the long milk tube and filter out the mastitis clots, thus telling the milker that a cow with mastitis is present.

Teat dipping

After milking, teats should be dipped in an antiseptic solution to kill any bacteria that may have lodged on the teat from the teat cup or the milker's hands. Glycerine and lanolin are often mixed with the antiseptic to keep the skin of the teat supple and to reduce chapping. This care of the teats reduces the number of areas where bacteria can live and multiply. Dipping is normally carried out by means of a cup, but automatic sprays are available.

Milking plant maintenance

Any worn or badly maintained milking equipment can cause undue stress on the udder tissue, so increasing the risk of mastitis. The following is a list of items that should be checked, together with a description of the role they play in causing mastitis:

1. Check the vacuum level of the plant. An unstable vacuum causes milk to flood from one teat up to another, so risking cross-contamination of quarters.
2. Check the pulsation rate (the number of times the liner opens and closes each minute) and the pulsation ratio (the proportion of each cycle for which the liner remains open or closed). Faulty pulsation can cause damage to the teat end and erosion of the sphincter muscle, allowing easy access for infectious organisms.
3. Worn or faulty teat cup liners will slip down the teat, letting in air. This alters the pressure in the cluster and can cause milk to be forced back against the end of the teat and injected back into the udder.
4. Automatic cluster removers (ACRs) (see Fig. 16.1) have helped to reduce the incidence of mastitis caused by overmilking. However, they must be correctly set. Ensure the air bleeds are clear, as a blockage can cause them to stay on too long and lead to severe bruising of the teats.
5. All rubber tubes must be in good condition, as even small holes will admit air and upset the milking pressure. Check that the air bleed on each claw piece is clear. This avoids milk flooding all the teat cups.
6. All the milking equipment must be thoroughly washed at the end of each milking. There are a number of methods of doing this, but normally it takes the form of a cold water rinse followed by a hot water wash with chemicals added to aid the process. The manufacturers of cleaning chemicals will state what temperature the cleaning water should be during circulation. The temperature of the water should be checked at the end of the cleaning cycle to ensure that an adequate temperature has been maintained.

   The Milk Marketing Board and milking plant manufacturers offer a milking plant testing service where, for a fee, specialist personnel carry out a full examination of the milking plant. A full report is provided and adjustments made if necessary. This is best done on a six-monthly or annual basis and should be regarded as an essential element of the farm's mastitis prevention programme.

Mastitis records

All mastitis cases should be recorded. Apart from the legal requirement of recording all drugs used on the farm, the mastitis record will give an indication of repeat treatments, so that carriers can be identified. The records can be analysed to give an annual percentage figure of mastitis cases and to identify persistent offenders.

FIG. 16.1 A milking parlour that features computerized feeding, milk recording and an automatic cluster removal system.

Early detection and treatment

It is essential that any cases of mastitis are detected before they develop into 'serious' cases and that a suitable antibiotic is administered for the specified time, with the contaminated milk being discarded. The type of antibiotic will be a matter for the veterinary surgeon but will normally be determined by bacteriological examination of mastitis samples. The samples are taken to a laboratory, the bacteria are grown on plates and antibiotic sensitivity tests are carried out. Treatment is by means of an infusion of antibiotic directly into the teat canal.

Dry cow therapy

Dry cow therapy involves the infusion of long-acting antibiotics into all four teats of the animal when she is dried off.

The object is to treat mastitis carriers and sub-clinical cases in order to reduce the spread from cow to cow. Some veterinary surgeons have suggested that dry cow therapy reduces bacteria in the udder to such a low level that the cow has less resistance to other forms of udder infections such as *Escherichia coli* mastitis, but this has not been proved.

TABLE 16.1 *Cell count payment system (correct January 1990).*

Cell count* ('000 per ml)	Price adjustment (pence per litre)
Up to 400	+0.2p
401–700	nil
701–1000	−0.2p
More than 1000	−0.4p

Source: MMB.
* Based on a rolling, three-month, geometric average.

Culling carrier cows

Cows that repeatedly contract mastitis may be culled in an effort to reduce cow-to-cow infection. This may seem a rather drastic step, but in terms of money spent on antibiotics, discarded milk and reduced lactation yields, a cow that frequently has mastitis is very expensive.

Cell count

It is possible to identify carrier cows by use of a technique known as the cell count. When mastitis-causing bacteria attack the udder tissue there is an upsurge in the number of white blood cells, as they are mobilized to try to defend the udder. The infection also causes a number of normal udder cells to be shed. It is these white blood cells and udder cells which are used to determine the cell count, the numbers present being an indication of the severity of the mastitis infection.

Cell counts are shown as thousands of cells per millilitre of milk, a figure of 250 (i.e. 250,000 cells per ml) or below being excellent.

From October 1991 the cell count figure for a herd will be determined on a weekly basis and be linked to the producer's payment for that milk. Table 16.1 shows the cell count payment system.

Cell counts are normally done on a whole herd basis, but if a herd's cell count is continually high then individual cows can be sampled to identify the offenders. The Milk Marketing Board's Veterinary Laboratory at Worcester is able to carry out cell counts on a contract basis for herds that require specialist help.

Mastitis is not the only cause of high cell counts. An increase in the number of late lactation cows in the milking herd that naturally shed milk-producing cells can have an effect and if milk from one mastitis case gets into the bulk tank the cell count will rise.

Mastitis transmitted from the environment

There are two main causal organisms of this type of mastitis:

Streptococcus uberis—a bacterium naturally occurring in the mouth, the skin, the teats and the faeces of cows.
Escherichia coli—a bacterium found in the large intestine and the faeces of cows.

Mastitis caused by *S. uberis* is sensitive to penicillin and is usually less severe than the infection caused by *E. coli*. Mastitis caused by *E. coli* can have a huge range of symptoms, from a slightly inflamed quarter with stringy milk clots, to death within a very short time. Cows will sometimes lose the milk production from the affected quarter, but this will often return at the next lactation.

Prevention and control

The causal organisms of environmental mastitis are found in the buildings in which the cow lives and in the faeces which contaminate those buildings. Prevention must be aimed at keeping faeces and contaminated bedding away from the teat. It is interesting to note that there are fewer cases of *E. coli* mastitis in the summer, when the cows are out, than in the winter. The following is a list of possible sources of infection.

Cubicle passages

These should be scraped out at least twice a day to avoid cows carrying faeces on to the cubicle beds where it could come into contact with the udder.

Cubicle beds

The beds must be clean and dry. There are many types of cubicle base—from concrete to sand—and many forms of bedding—from chopped straw to shredded paper. Each has its own advantages and disadvantages. The overriding factor must be cow comfort. If the cow is comfortable in clean, dry, well-designed cubicles, it will not only reduce the stress on the animal but also the chances of faeces getting into contact with the teats.

Some veterinary surgeons advise that a small amount of slaked lime should be sprinkled on the bed. This acts as a disinfectant and also dries the bed. However, too much lime can cause soreness to sensitive teats.

Overcrowding

Overcrowding of the cows must be avoided, as it leads to faeces being splashed onto the teats, skin and coats as a large amount of faeces gets deposited in a small area. Cows must have adequate lounging area and the passageways must be wide enough to avoid crushing.

Calving boxes

It is essential that calving boxes be kept as clean as possible. Calving cows seem to be at most risk to *E. coli* mastitis due to contamination of the calving box by milk and faeces, a medium in which *E. coli* bacteria will multiply at an alarming rate.

Farms that have a very tight calving pattern (a lot of cows calving in a short period of time) have an increased risk of cows contracting *E. coli* mastitis in the calving box, as there is often no break from one cow to the next.

Parlour hygiene

During milking, the teat sphincter muscle is open, allowing bacteria to enter. The milker must make every effort to ensure that the teats are not contaminated with faeces when the units are put on.

Summer mastitis

The major causal organism of this disease is the bacterium *Corynebacterium pyogenes*. It affects cows and heifers, and occasionally very young heifers, out at pasture in the summer months of July and August. The illness is usually severe and death is not uncommon.

An affected animal will have a stiff-legged walk and one or more quarters will be hot and swollen and may be quite painful. If the quarter is milked, the milk will be thick and yellow with a foul smell.

It is thought that the biting fly *Hydrotaea irritans* (head fly) is responsible for transmitting the disease from animal to animal, but there is no proof that the fly is responsible for the disease developing in the first place. The flies are attracted to the underside of the cow and the ends of the teats where they can be seen gathering in large numbers. One of the early signs to watch out for is flies on a teat that is slightly swollen.

Prevention and control

1. Dry cow therapy as used for the prevention of 'normal' mastitis is useful, as the long-acting antibiotic will help prevent infection. This is only useful for cows as heifers will not have received any dry cow therapy. However, heifers can be protected by smearing the outside of the teat sphincter with a long-acting antibiotic after the teat has been cleaned.
2. As flies are responsible for transmission of the bacteria, some form of fly control is essential. Long-acting insecticides can be sprayed on to the animals, not forgetting that it is the underside that requires most attention, or a plastic ear tag impregnated with insecticide can be used. There are several of these tags on the market. They work by the animal grooming itself and wiping the tag across its coat. As it does so, some insecticide gets dissolved in the coat's natural oils which then cover the body. As the coating of hair and body oils does not cover the teats and udder to the same extent as it does the shoulders, this method of fly control has limited use for summer mastitis.
3. Summer mastitis often occurs around woodland and boggy areas where flies are present in great numbers. If these areas can be avoided from July to September the incidence of the disease can be reduced.
4. Early detection is very difficult. Once the animal shows obvious symptoms, the damage has been done and the quarter is normally lost completely. However, early detection will reduce the severity of the disease and, if the affected animals are isolated, reduce the chances of other animals becoming infected.

Total bacteria count of milk

The total bacteria count (TBC) is a measure of the number of bacteria in a milk sample. These could have come from either the cow or the milking equipment. So unlike cell counts, which are a measure of the number of white blood cells in the milk that come from the cow only, TBC takes into account the cleanliness of the milking equipment and thc milking routine.

Cracked rubber tubes, inadequate temperature of the washing water, inefficient cleaning of the bulk tank and faeces on the teats are just some of the factors that can lead to a rise in the TBC.

In 1982 the Milk Marketing Board introduced a system whereby the TBC was taken into account for the payment of the milk. There is an additional payment to the farmer if the milk is of top hygienic quality but money is deducted as the TBC rises (Table 16.2).

This obviously encourages the farmer to produce 'clean' milk, the hidden bonus being that the measures taken to reduce the TBC will also help to reduce the incidence of mastitis in the herd as a whole.

Antibiotic contamination of milk

Any farmer selling milk to the Milk Marketing Board (MMB) signs a contract, part of which says that the farmer will not sell milk contaminated with antibiotics and other 'inhibitory' substances. The reasons for this are that antibiotics will destroy the bacteria in milk used for cheese and yoghurt manufacture and that some people are allergic to antibiotics, especially penicillin.

Antibiotics can get into the bulk tank for a number of reasons, a common one being poor communication between milking personnel. Any cow whose milk is to be withheld for any reason must be clearly marked. The milkers must know what the mark is, what it means and act accordingly. Confusion occurs when relief milkers take over and the normal herdsperson has not left instructions. It is important to check the persistence of antibiotic preparations and to record each cow when treated. Dry cow therapy can also be a problem. The MMB contract states that milk should be withheld from a cow for the first four days after calving. This gives a chance for antibiotic residues to clear. However, if a cow treated with long-acting antibiotic calves earlier than anticipated, the four days may not be long enough. Accidental contamination can

TABLE 16.2 *Total bacterial count payment system (correct January 1990).*

Band	Average number of microorganisms per ml	Milk price adjustment (pence per litre)
A	Up to and including 20000	+0.230
B	More than 20000 and including 100000	nil
C	More than 100000	−1.500 (if there has been no deduction in the previous six months) −6.000 (if there has been a deduction of 1.500 in the last six months) −10.000 (if there has been a deduction of 6.000 or 10.000 in the previous six months)

The placing of a supply into a band will be based on the results of at least two tests carried out on the month's milk supply.

For Band A the average must be in Band A or all results except one must be in Band A.

For Band C the average must be in Band C and two or more results must be in that band.

In every other case the supply will be classified in Band B.

Source: MMB.

happen if, for example, the milker switches a valve the wrong way, milk may be sent to the bulk tank instead of being discarded. Cows under treatment should be milked into a separate bucket. If they are milked into a jar it must be flushed out with cold water before the next cow is milked.

Ideally, treated cows should be milked last to avoid any chance of contamination, but normally this is not practical.

All farms have a sample of milk taken and tested once a week. If antibiotics or other substances are found in the milk, the farmer is paid a very low price for that consignment. Repeated failures can put the contract in jeopardy. If a farmer suspects that antibiotic has got into the milk, the MMB must be informed before the milk is collected. The milk will be tested and, if found to be contaminated, a separate tanker will be sent to collect the milk.

From October 1990 the antibiotic sensitivity test has been made a lot stricter. Previously, all milk was deemed contaminated with antibiotics if more than 0.01 international units (iu) per ml of milk could be detected. From October 1990 the figure has been 0.005 iu per ml.

Mastitis in the sow

Mastitis in the sow is often referred to as 'farrowing fever' and is associated with a complex condition known as metritis, mastitis and agalactia syndrome (MMA). It occurs within two or three days of farrowing and typically the sow has a high temperature (42 °C), is off her food, the skin is flushed and part of the udder is hot and hard. There is little or no milk in the udder and teats—agalactia—and there is inflammation of the uterus, occasionally with a discharge from the vulva—metritis.

The condition has a dramatic effect on the piglets. They get little food and show distinct signs of malnutrition such as sunken eyes and 'razor' backs and they are excessively noisy. They quickly lose the will to live.

In order to avoid serious losses to an affected litter the piglets must be fostered on to other sows or taught how to drink sow milk replacer (specially formulated powdered milk) from shallow bowls or troughs. Treatment for the sow involves the use of broad spectrum antibiotics and injections of oxytocin to stimulate milk let-down. Prompt treatment (2–3 days) is necessary for the sow to make a quick recovery. The causal organisms seem to be *Escherichia coli* and *Corynebacterium pyogenes*, but what makes these naturally occurring bacteria suddenly cause this disease is not well understood.

There are a number of factors involved, but it may only be one that tips the balance in favour of the disease. Sows under stress are certainly more prone to infection, so sows should be moved to their farrowing quarters early (at least a week) to give them time to get used to the pen. *E. coli* is found in faeces, so poor hygiene in the farrowing quarters certainly plays a part. Sows should be moved into farrowing pens that have been cleaned, disinfected, and are dry and warm.

MMA is often associated with high stocking rates and the subsequent pressure on space. If a unit has a constant problem, then letting sows farrow in loose boxes and other simple types of accommodation can often help. The piglet losses incurred by the increased incidence of overlying can often be less than those incurred by the disease itself.

Mastitis in the ewe

The causal organisms of mastitis in the ewe are *Corynebacterium pyogenes*, *Staphylococcus aureus* and *Pasteurella haemolytica*. These organisms occur naturally on the skin, so it is thought that the predisposing cause may be a cut to the teat or even a sudden spell of cold weather.

Acutely affected ewes have a stiff-legged walk and their lambs are weak and hungry. Part of the udder turns black and is cold to the touch. If the teat is drawn, it reveals a clotted blood-stained fluid. The whole of the udder or the affected part may degenerate and come away completely. Less acute forms of the disease are not so easily diagnosed but are quite common. They come to light at culling time when the udders are inspected and may be found to be hard and swollen or small with fibrous lumps inside.

Prevention is difficult, as the cause is often unclear, but poor hygiene at lambing time, especially if lambing takes place indoors, can lead to an increase in the disease.

Ewes kept in upland areas do not have docked tails. The tail hangs down covering the rear of the udder and so helps to stop chilling.

Chapter 17

Diseases of the young animal

Young animals, especially new-born ones, are more vulnerable to infections than older animals. This is largely owing to the absence of any active immunity in the animal's blood stream, but another reason is that the immune system takes time (two to three weeks) to develop. Also, animals are often born into a 'dirty' environment and so meet a heavy bacterial challenge very early in life.

By following a few basic rules of stockmanship a lot of these problems can be overcome. The first and most important rule is that every new-born animal must have access to colostrum, preferably its own mother's, within the first few hours of life. Colostrum contains antibodies to many of the diseases that the mother has been exposed to in her surroundings, these diseases being the most likely ones that the young animal will meet. For the young animal this acquired immunity will last until its own immune system has developed sufficiently to produce its own antibodies. The antibodies in the colostrum do not last for long and the ability of the animal's gut to absorb them soon deteriorates, so it is important that young animals get colostrum within the first few hours of life.

The place where the mother gives birth must be clean. This is relatively easy to achieve as far as calving boxes or farrowing pens are concerned, as they can be cleaned and disinfected after each animal. It is not so easy for sheep, as a large number of ewes may give birth in a short period of time in a confined area (Fig. 17.1), the later ones being exposed to the bacteria and viruses of the earlier lambers. There is no easy way round this except to ensure that each lamb gets colostrum and that the normal hygiene precautions, such as ensuring all lambing equipment is clean and disinfected, are strictly followed.

Newly born animals need to be warm. This is most obvious in piglets—small animals with little or no body hair, a large surface area relative to their weight and a poorly developed heat-regulating mechanism. However, calves also need a warm, dry bed and a place that is sheltered from draughts and the worst of the weather. Even lambs will suffer if their fleece is continually soaked through with rain soon after birth. Some farms have adopted a system of using lamb coats, a loose fitting plastic cover, to protect weakly lambs from heavy rainfall.

If an outbreak of disease is suspected in an individual or a group of animals, it, or

FIG. 17.1 Lambing pens constructed under an open barn. When lambing is finished they are dismantled, the hurdles cleaned and the bedding removed. This provides a clean lambing area for the following year.

they, should be isolated as far as possible from other stock in order to reduce the spread of the disease. Simple precautions such as not using the same feeding equipment for sick and healthy animals, foot dips and feeding sick animals last will also help to stop disease from spreading.

Most infections will enter young animals through the mouth, but infection may also occur through the navel chord. Many stockpeople routinely 'dress' the navel chord with an antiseptic such as iodine as soon as the animal is born, in order to minimize diseases entering via this route.

Diseases

This section will consider some of the more common diseases that affect young animals. The two most important are:

Scouring (digestive diseases), which may be caused by a number of organisms including salmonella.
Pneumonia.

Other diseases that need to be considered in relation to young animals are:

Joint-ill (calves/lambs/piglets).
Navel-ill (calves/lambs/piglets).
Calf diphtheria.
Atrophic rhinitis (pigs).
Meningitis (calves/lambs/pigs).
Greasy pig disease.

Scouring

Scouring is the most common disease of young animals and is the disease that causes the most losses, both physical (death) and economic (poor or stunted growth). However, scouring is only the symptom of a number of infections and problems that cause, among other things, diarrhoea.

Before considering these infections and their causal organisms, it is worth discussing what scouring actually is and what damage it does to the small intestine.

The inside of the small intestine is covered with millions of finger-like projections called villi (*see* page 16). These are not only responsible for absorbing digested food (amino acids, sugars, etc.) into the bloodstream, but also for releasing water from the blood into the intestines, as part of the process of digestion.

Most of this water is then reabsorbed into the body as it passes down into the large intestine, resulting in semi-solid faeces being voided by the animal. If the animal picks up an infection in the intestines, this delicate balance is upset and the animal begins to lose more water in the faeces than it is taking in—scouring. The result is dehydration. Bacteria attach themselves to the villi and stimulate more water to be released into the intestine, whilst viruses, on the other hand, attack and destroy the villi and also reduce the amount of water re-absorbed into the body.

Excessive water loss is not the only problem the animal has to contend with, as nutrients and salts that should have been absorbed, if the villi were not damaged, are also voided with the faeces.

Treatment of scours

The following treatments are of a very general nature, more specific ones being dealt with under the actual diseases. Stockpeople should note that scouring animals can dehydrate and die very quickly, the smaller the animal the quicker it will dehydrate.

1. Provide fluids to compensate for the loss of water. The water should be clean and, if offered in a bucket or tub, should be changed several times a day.
2. Some animals are so weak that they cannot drink. There are stomach tubing kits available for calves, lambs and piglets where fluid can be passed via a tube down the oesophagus, straight to the stomach. If an animal is very weak this method is preferable to drenching.
3. There are several commercial preparations available that will replace salts and nutrients that have been lost. These can be administered via the water supply or directly through a stomach tube.
4. Specific treatments as recommended by a veterinary surgeon. These may include antibiotic therapy, but the treatment depends on the causal organism.

Prevention of scours

1. Ensure that all feeding utensils are cleaned and disinfected after each feed.
2. Separate scouring animals from healthy ones.
3. Thoroughly clean and disinfect all animal buildings when the animals have moved out.

4. Make sure that the living area is clean, dry and free from draughts.
5. In the case of calves fed on milk substitute, ensure that it is fed at the correct temperature, at the correct strength, is free from clumps of unmixed powder and is fed at the same time each day.
6. It is possible to vaccinate against some causal organisms of scours. Depending on the type of vaccine, it may be administered to the dam to pass immunity to the offspring in the colostrum or given to the young animal itself. If it is given to the young animals, it may take up to three weeks for the body to gain full immunity to the disease.

Specific causes of scours

Escherichia coli

The *E. coli* bacterium is responsible for causing scouring in calves, piglets and lambs. There are two definite forms of the disease. The first occurs within the first few days of life and causes severe watery diarrhoea and a high temperature. The second occurs between two and seven weeks of age and causes a white-coloured scour.

Both forms of the disease are responsible for a high percentage of deaths and 'poor doers' in calves, piglets and lambs. Specific symptoms vary between the different species, making it too complex a disease to be covered fully in this book. Any stockperson suspecting a case (or cases) of *E. coli* scours should consult a veterinary surgeon.

Treatment and prevention is a combination of antibiotics or an antiserum in the first instance and possibly vaccination as a long-term preventive measure.

Salmonellosis

Salmonellosis is a bacterial disease of many species and affects animals of all ages. It is caused by infection with organisms of the salmonella group. There are literally hundreds of different types of salmonella bacteria; a few of the most important are *S. dublin* and *S. typhimurium* which affect cattle, pigs and sheep, *S. agona* which affects sheep and *S. cholerae-suis* which affects pigs.

A wide range of symptoms include high temperature, severe scouring, loss of body condition and, in the case of lactating females, loss of milk yield and possibly abortion. The animal may pass blood-stained faeces and even shreds or lumps of the mucous membrane from the intestine.

Some animals can be carriers of the disease but show no symptoms, the bacteria being excreted in the faeces for others in the group to catch.

As with *E. coli* scours, this is a complex disease requiring treatment under veterinary supervision.

Salmonellosis is an important cause of food poisoning in man (*see* page 187) and as such it is subject to a Zoonoses Order. This requires any outbreaks of salmonella on a farm to be reported to the appropriate authorities in case the disease is passed on to humans. There is specific legislation covering outbreaks of salmonellosis in poultry.

Treatment is by the careful use of antibiotics. Misuse of these drugs can lead to resistant strains of the bacteria and an increase in the number of animals becoming carriers.

Prevention is possible by a programme of vaccinations.

Coccidiosis

Coccidiosis is caused by a protozoan parasite *Eimeria zuernii* and affects calves, piglets and lambs. The parasite invades the cells in the wall of the large intestines where it reproduces, eventually producing eggs or 'oocysts'. These oocysts are passed out in the faeces and will infect other animals via the mouth.

Infected animals will be scouring, often with blood tainting the faeces. Calves and lambs will be straining to pass faeces with only small amounts being forced out. Piglets have a watery, yellow-coloured diarrhoea that has a disgusting smell.

Diagnosis of the disease is by laboratory examination of the faeces in order to identify any oocysts.

Treatment is by dosing with specific anti-coccidiosis drugs as prescribed by a veterinary surgeon.

The oocyst has a very thick wall and can live in buildings and bedding for many months. Long-term prevention is dependent on improved hygiene and thorough cleaning and pressure washing of pens before new animals are bought in. Oocysts are resistant to many ordinary disinfectants, so a specific anticoccidial disinfectant should be used.

Rotavirus infection

This viral infection is seen in calves up to a week old and in new-born piglets. (It has also been isolated in scouring lambs, but as yet is not recognized as a serious problem.)

The virus causes a watery diarrhoea which quickly leads to dehydration of the animal. This is because the virus destroys cells lining the villi of the small intestine, preventing the reabsorption of water. The villi can be permanently damaged and this may result in stunted growth due to the animal's inability to absorb the products of digestion.

It is more common in calves born from heifers and litters from gilts, as these young mothers may not have built up the immunity that the older animals in the herd have.

There is no specific treatment, although fluid replacement therapy is vital.

A vaccine is available to protect calves from the disease.

Transmissible gastroenteritis (TGE)

This is a viral disease of pigs causing huge losses in piglets aged three weeks and under. It can also affect pigs of any age, but with a lower mortality rate.

The symptoms are a yellow watery diarrhoea and vomiting which spreads throughout the herd at an alarming rate. Lactating sows may have little or no milk, whilst growing pigs may have only very slight symptoms.

Immunity will build up in the herd, but it does take time. It is important that new breeding stock is exposed to the disease in order to gain some immunity.

In some countries a vaccine is available, but this is not yet so in the United Kingdom.

Swine dysentery

Swine dysentery is a disease of pigs caused by the bacterium *Treponema hyodysenteriae*. It normally affects growing pigs, but pigs of all ages can be affected.

The symptoms include blood-streaked diarrhoea, mucus in the faeces and pigs standing with an arched back kicking at their stomachs, probably because of the pain. As the disease progresses, affected pigs gradually lose body condition, although they will usually drink quite happily.

The disease is passed rapidly around the unit by carrier pigs and infected faeces.

It is possible to medicate the water supply in order to treat batches, but, at the same time, it is important to ensure that no faeces are spread about the unit, transmitting the infection to non-infected pigs.

Prevention is by ensuring that the pigs are separated from their faeces and that strict hygiene precautions are used.

As the disease can be brought in by carrier pigs, any replacement stock should be obtained from a non-infected source.

There are two other common causes of scouring that are discussed under their more appropriate chapters. These are lamb dysentery (page 159) and parasitic gastroenteritis (page 180).

Pneumonia

Pneumonia is a disease that mainly affects young animals, although older ones are certainly not immune. In its broadest sense it may be defined as inflammation of the tissues of the lungs.

If the various types of pneumonia are classified according to their cause, there are six main types:

Viral pneumonia.
Mycoplasmal pneumonia.
Bacterial pneumonia.
Parasitic pneumonia.
Mycotic (fungal) pneumonia.
Non-infective (mechanically caused) pneumonia.

Of these six types, viral, mycoplasmal and bacterial are the most common. Parasitic pneumonia is discussed in Chapter 23, whilst mycotic pneumonia is fairly rare.

It is worth mentioning briefly non-infective (mechanically caused) pneumonia. This can be brought about by poor drenching techniques and by treatments administered by means of a stomach tube which mistakenly goes into the trachea and not the oesophagus. Pneumonia may also be caused by an animal inhaling its own vomit.

Symptoms of pneumonia

As pneumonia is a complex disease and more than one causal organism may be involved, it is difficult to describe the specific symptoms of the various types. However, there are some symptoms that are common to cattle, pigs and sheep.

The first sign of pneumonia, especially in calves, is a slight watery discharge from one or both eyes which will leave a tear-stain on the face. In all species a dry cough will develop, associated with a faster respiration rate. Severely affected animals will have an arched back, head down and be struggling to get their breath. Animals with pneumonia will normally have a raised temperature and be off their food.

In very severe outbreaks some animals may die before any coughing is heard.

Pneumonia in pigs can lead to severe economic loss, especially if it is caused by the bacterium *Haemophilus pleuropneumoniae*. This has been seen in the United Kingdom since the early 1980s and affects pigs when they have been weaned. The pigs lose their appetite, show a marked reduction in food conversion ratio and growth rate and can have severe coughs with breathing problems, eventually leading to death.

The problem with this type of pneumonia is that there are several different sub-groups of the causal organism, so no general vaccine is available. However, it is possible to get a specific vaccine specially made to combat the particular strain on a farm. This is done through a veterinary surgeon.

Prevention of pneumonia

Animals will go down with pneumonia for a number of reasons—they may have no immunity, they may be housed in adverse environmental conditions or the level of infection within a building is extremely high.

Whatever the reason, there are certain steps that a stockperson can take that will help to reduce the risk.

1. Provide adequate ventilation. As the causal organisms are often carried in the atmosphere, a constant supply of clean, fresh air will help to reduce the number of organisms in the atmosphere. This is one of the great challenges for designers of pig buildings: to provide fresh air and yet still maintain a desirable living temperature for the animals.
2. Reduce the humidity of the air in the building. Pneumonia-causing organisms will live longer in air that has a high humidity than they will in air with a low humidity. This is why the incidence of pneumonia will rise on foggy, still days. Lowering the humidity is achieved by good ventilation and by providing bedding to soak up wet urine.
3. Do not overstock the building. Overcrowded animals will not only pass diseases from one to another, but will also be under stress and so be more susceptible to disease.
4. Provide a dry and warm living environment. This will reduce stress on the animal to a minimum.
5. Avoid mixing animals of different ages. Older animals may have quite a high level of immunity but may be excreting causal organisms into the atmosphere. If they are mixed with younger animals that have a lower immunity, the young animals will succumb to the disease.
6. Ensure an adequate level of immunity. This is done firstly by making sure that each new-born animal receives colostrum, and secondly, by a programme of vaccinations. There are a number of vaccines available, some aimed at a range of organisms, others targeted at one specific organism. Vaccines should be chosen and administered under veterinary supervision.

Other diseases of young animals

Navel-ill and joint-ill

Navel-ill is a bacterial infection of the navel that may progress to joint-ill, a bacterial infection of the joints.

The disease affects calves, lambs and, to some extent, piglets. A number of bacteria, the commonest of which are streptococci, staphylococci and *Escherichia coli*, enter the wet navel cord as soon as the animal is born. They will then form abscesses in the navel which will be hot and hard to the touch and have a foul smell. If this is not treated early, the bacteria may enter the blood stream and settle out in the animal's joints, the most common ones affected being the knee, hock, hip, shoulder and stifle joints. Abscesses form deep inside the joints, they swell up and the animal either walks very stiffly or is reluctant to move. Other symptoms include a raised temperature and an unwillingness to suckle or eat.

In piglets navel-ill often goes unnoticed, probably because the animals are so small and so close to the ground that the navel is difficult to see. However, joint-ill is fairly common, with the result that affected piglets are pushed away from the udder at feeding time because of their inability to walk properly.

Calves and lambs can be affected quite severely by both navel-ill and joint-ill, leading to death in some cases.

Treatment involves the use of antibiotics and, occasionally, the navel or affected joints may have to be cut open and drained of pus. There are a number of measures that can be taken in order to prevent the disease.

1. Make sure that the animal is born into a clean environment. This means cleaning and disinfecting calving boxes and farrowing pens and placing lambing pens on a new site each year. Any equipment used at births, such as calving ropes, scalpels and scissors, must be clean.
2. All surgical equipment, for example piglet tooth clippers and ear notching pliers and anything else that may make a wound—ear tags for calves and lambs for example, should be clean and disinfected as wounds can provide entry for the bacteria.
3. The most common preventive measure is to treat the navel of each new-born animal with an antiseptic to seal it and kill any bacteria present. Iodine in dip or spray form is widely used for this purpose.

Calf diphtheria

This is a severe infection of the soft tissues in the mouth and, sometimes, nose, of calves aged about six weeks, caused by the bacterium *Fusiformis necrophorus*. The bacterium enters the skin through a cut or abrasion in the calf's mouth and quickly forms a pus-covered ulcer, often inside the cheek of the calf. Sometimes the tongue can be affected, leading to drooling, frothing at the mouth and difficulty in swallowing food.

Treatment is by antibiotic injection and a suitable diet, as the calf may need liquid feed until it can eat properly.

Dirty water troughs or bowls and dirty feed troughs can cause infection. Coarse hay that contains thistles may lead to damage to the inside of the mouth.

It should be noted that although this disease has the same name as one that affects humans, the causal organism is different.

Atrophic rhinitis

Atrophic rhinitis is a respiratory disease of pigs where a number of organisms (principally the bacteria *Pasteurella multocida*, *Bordetella bronchiseptica* and the Rhinitis virus inclusion body) attack and destroy the membranes and turbinate bones of the nose. Figure 17.2 shows a cross-section of (i) a normal nose, (ii) some destruction of the turbinate bones, and (iii) severe destruction of the turbinate bones.

The disease is contracted by young piglets from other piglets or carrier sows, but the symptoms are not apparent until the pig is at the growing stage (10 weeks plus). The symptoms include sneezing, tear-stained faces, inflammation of the snout and, in severe cases, bleeding from the nose. As the pig gets older the snout may be twisted or shortened owing to the degeneration of the supporting turbinate bones.

The disease can cause huge economic losses in the form of reduced growth rates and reduced food conversion ratio. Also, because the function of the turbinate bones is to warm the air that is breathed in and to filter out dust from that air, destruction of the

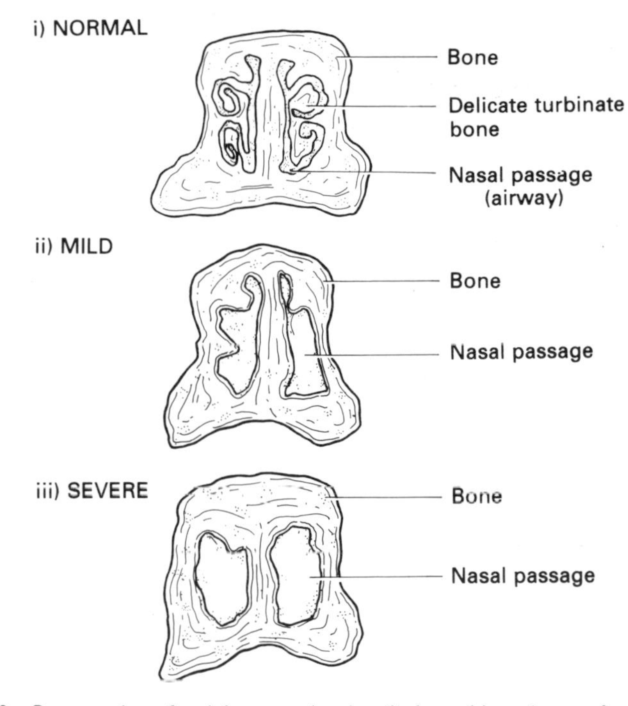

FIG. 17.2 Cross-section of a pig's snout showing (i) the turbinate bones of a normal pig; (ii) the turbinate bones mildly affected by rhinitis; (iii) the turbinate bones severely affected by rhinitis.

turbinate bones makes the pig more prone to respiratory infections, including pneumonia.

It is a particularly difficult disease to treat effectively as it is closely linked with many aspects of the pig's environment such as ventilation, the amount of dust in the atmosphere and stocking rate. Treatment involves the use of medications in the ration linked to improvements in management and the environment. A vaccine is available as a preventive measure.

Some pig farms that have been badly affected by rhinitis have taken the decision to depopulate the unit completely and restock using rhinitis-free breeding stock. This might seem like drastic action to take, but it could be more cost effective in the long term.

It is possible to get a measure of the degree of infection in a herd by regular assessment of cross-sections of snouts of pigs sent for slaughter.

Meningitis

The term meningitis means inflammation of the membranes surrounding the brain or the spinal cord (the meninges). There can be a number of causes—in calves it can be caused by streptococci and *Escherichia coli* bacteria, in lambs by a pasteurella bacterium and in pigs by *Streptococcus suis* (see Zoonoses, page 188). Viruses can also cause meningitis.

As would be expected of a disease that affects the brain, the symptoms are of excitement and uncoordinated behaviour. Animals will often walk round in circles bumping into objects and may appear to be blind. Pigs collapse and make characteristic paddling movements with their legs, often accompanied by a high-pitched squeal. In all affected animals the body temperature is above normal.

Treatment is by use of antibiotics, usually at a very high dose-rate. If the disease is spotted and treated in its early stages the animal can be saved and show no after-effects. However, if the disease is allowed to progress, meningitis will cause death or may leave the animal with permanent brain damage, showing up as uncoordinated movement or strange behaviour.

Greasy pig disease

This is a bacterial infection of the skin caused by *Staphylococcus hyicus*. It gains entry through cuts and scratches and causes a yellow, dirty, greasy appearance to the skin, often around the neck and face, but it can cover the whole body.

Any age of pig can be affected, but it is most commonly seen in recently weaned piglets as they are prone to fighting when mixed, resulting in scratches to the ears, neck and flanks. Biting flies that live in the farrowing pens or weaner area can also transmit the disease.

All the pigs in the group affected should be sprayed with a skin cream prescribed by a veterinary surgeon. Treatment with antibiotics is possible if the disease is caught in its early stages.

Chapter 18

Diseases that affect the reproductive system or cause abortion

The ability of farm animals to produce offspring on a regular basis is an important factor in the profitability of a livestock enterprise and maintaining that fertility is a task that occupies a large part of the stockperson's working day.

The factors that affect the fertility of farm animals cover a huge area, including nutrition, housing, daylight hours, age and many more. The object of this chapter is to look at just one of these factors, namely disease.

The diseases discussed here are only the main ones. The chapter covers only the 'normal' diseases that affect the reproductive system and does not explore the more complex issues of, for example, whether the housing, nutrition or even management are contributory factors to the illness of the animal.

Some of the diseases listed below have abortion of the fetus as one of the symptoms. It should be noted that any cases of abortion should be reported to a veterinary surgeon and that cows that have aborted should not be offered for sale until two months after they aborted. This is a preventive measure to help stop the spread of brucellosis, a notifiable disease.

Brucellosis and Aujeszky's disease

Both brucellosis in cattle and Aujeszky's disease in pigs may result in infertility and abortions, but, as they are both notifiable diseases, they are dealt with in Chapter 11.

Trichomoniasis

Trichomoniasis is caused by a microscopic parasite—*Trichomonas fetus*. It affects cattle and can cause serious economic losses due to infertility problems.

A bull becomes infected from serving an infected cow. The parasites then live in the skin that covers the penis and thus the bull may go on to infect other cows that he serves. The bull usually shows no symptoms of infection, but cows may have in holding to service, have a discharge from the vulva and may abort. Diagnosis is by laboratory test of either mucus from suspected cows or swabs from the prepuce of the bull.

The widespread use of artificial insemination has been a major factor in halting the spread of this disease between herds.

Preventive measures include identification and treatment of the disease and close management of services.

Campylobacter infection (vibriosis)

Both cattle and sheep can be affected by *Campylobacter*.

Cattle

In cattle, *Campylobacter fetus* can be an important cause of infertility, resulting in serious economic loss. It is a sexually transmitted disease that is spread by natural service. Bull semen used in artificial insemination is screened to prevent the spread of the disease.

Although the bull looks normal, the cow may show signs of discharge and may abort. Infertility is usually temporary.

Prevention and control is by resting infected bulls, good service management, washing the bull's prepuce with antibiotics and using AI.

Sheep

The disease in sheep is not spread sexually. It is passed on by a ewe eating grass that has been contaminated by womb discharges of infected sheep.

It is very contagious and may cause many abortions in the flock. Diagnosis is by laboratory examination of an aborted fetus or placenta.

Vaccination is possible but is expensive. There is no effective treatment, although flocks do seem to build up immunity to the disease.

Enzootic abortion in ewes

Enzootic abortion in ewes is caused by a micro-organism of the *Chlamydia* family. It is spread mainly between ewes at lambing time, but can also be sexually transmitted from ram to ewe. Lambs can be infected with the organism, which will remain latent in the lamb until it is itself pregnant, when abortions may occur.

If the placenta of infected ewes is examined it will have the appearance of being thickened and is covered in spots of pus.

An affected flock will rapidly build up immunity to the disease and a vaccine is available to be used under veterinary advice.

A high stocking density of ewes at lambing time will aid the rapid spread of the disease.

It should be noted that pregnant women should avoid any contact with ewes at lambing time as *Chlamydia* can also cause abortions in humans.

Parvovirus infection of pigs (SMEDI)

Parvovirus infection in pigs produces the classic symptoms of the SMEDI syndrome, that is stillbirths, mummification, embryonic death and infertility. It affects breeding sows and results in small litter numbers, mummified fetuses and possible problems of the sow failing to hold to service.

The virus is transmitted by contact with infected dung and fetal membranes.

It is a difficult disease to detect unless careful analysis is made of the farrowing and breeding records, although laboratory analysis of a blood sample should confirm if the virus is present.

A vaccine is available to protect breeding animals (including boars), but the timing of its administration is very important and must be strictly followed. It is also important to ensure that all new stock are vaccinated.

Leptospirosis (see also Chapter 24, Zoonoses)

Leptospirosis affects cattle, pigs and man.

Cattle

Leptospirosis in cattle is caused by *Leptospira interrogans* serovar. *hardjo* and is sometimes referred to as 'flabby bag' or 'milk drop'. This is because one of the very noticeable symptoms in dairy cows is a sudden drop in milk production with the milk having a slightly thicker texture than normal. This may be followed some time later by abortion.

It may be that the loss of production goes unnoticed and the only symptom will be the abortion. Blood sample analysis will confirm the presence of leptospirosis.

The organism is present in the urine of infected cows. It is passed to others when this urine gets into their eyes, mouths or cuts on the skin surface.

Antibiotics can be used to treat affected animals.

A vaccine is available in order to provide long-term immunity for cows and heifers.

Pigs

Pigs may be affected by a number of types of leptospirosis organisms. The commonest type is *Leptospira interrogans* serovar. *pomona* which causes abortions and stillbirths. When a pig herd is first infected, an 'abortion storm' may occur, with many sows aborting. The incidence then falls as the herd gains some immunity to the disease.

Antibodies may be used under veterinary supervision, but a licensed vaccine is not yet available.

Leptospira interrogans serovar. *icterohaemorrhagice*, the organism that causes Weil's disease in man, may occasionally infect pigs, especially outdoor herds that gain access to contaminated water.

Metritis

Metritis is a term that means inflammation of the uterus. It affects cattle, pigs and sheep. In cattle it is often known as 'whites', due to the creamy white discharge that is seen coming from the vulva. It is caused by a number of bacteria and so will respond to antibiotic and pessary treatment. Acute cases can occur especially after a difficult or assisted birth when infection may have been introduced accidently into the reproductive tract.

Metritis in sheep is often caused by one of the *Clostridium* type of bacteria, so vaccinating against clostridial disease (*see* page 158) is a good preventive measure.

Metritis may cause temporary or permanent infertility, depending on the severity of the infection.

Mycotic abortion

Mycotic abortion is a term used to describe abortions that are caused by fungi or, more accurately, fungal spores.

Mycotic abortions have occurred in cattle, pigs and sheep, with cattle being the major group affected.

The fungus *Aspergillus* has been identified as a common cause of mycotic abortion. It grows on mouldy hay, silage, wet beet pulp and brewer's grains.

Abortion problems have occurred in pig herds where mouldy food has been fed to pregnant sows. If the food 'sweats' and gets damp whilst in the bulk feed bin, it will get contaminated by growths of mould.

Chapter 19

Metabolic diseases, mineral and vitamin deficiencies, digestive disorders

The idea that a disease might be due to a deficiency is easily understood. The most obvious example is an animal in an emaciated condition due to under-nutrition. The words 'deficiency disease' indicate a disease resulting from an inadequate supply of some essential item in the animal's diet. On the other hand, a metabolic disease implies a deficiency not in the diet but a malfunction of the vital chemical processes within the animal's body. So, for example, milk fever is a metabolic disease caused by a deficiency of calcium in the blood, but it does not arise from a lack of calcium in either the food or the animal's body. Thus, metabolic diseases and deficiency diseases are not caused by bacteria, viruses or other organisms.

Metabolic diseases

Milk fever (hypocalcaemia)

This condition affects cows and sheep just before or just after calving or lambing (lambing sickness) and is due to a shortage of calcium in the blood. As it is more common in cattle, the symptoms, treatment and preventive measures described here relate to the cow.

Milk fever usually occurs within a few hours or days after calving. The calcium contained in the blood and intestines is not enough to meet the sudden demand of milk production in very early lactation, although there are plenty of reserves within the body in the form of calcium on the bone. Heifers are less prone to milk fever as the calcium on their bones is more available than that of older cows.

Symptoms

If the animal has just calved she may not be able to stand. If standing, she will stagger and quickly fall over, she will be quiet and have a lower temperature than

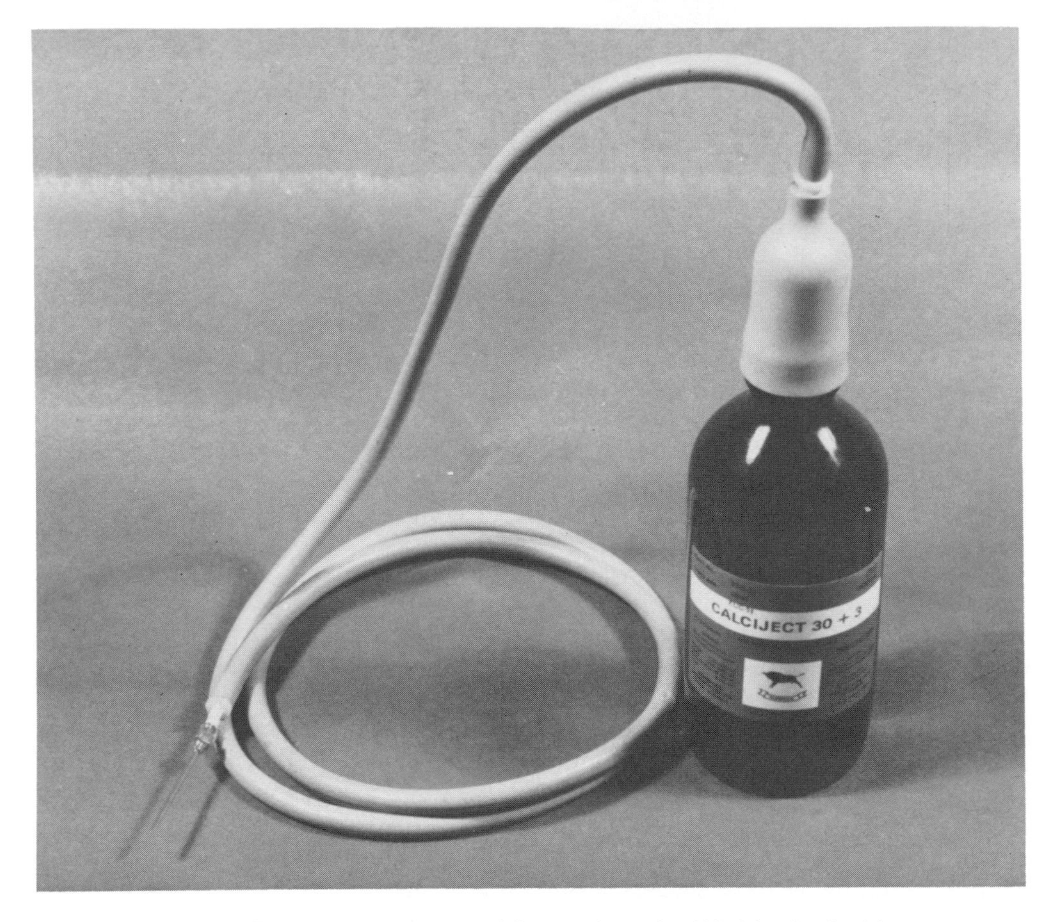

FIG. 19.1 The most common type of flutter valve and a 400ml bottle of calcium borogluconate used in the treatment of milk fever.

normal with a cold coat. As the milk fever gets worse the animal's movements get slower and cudding will stop altogether. If left untreated, she will roll onto her side preventing the gases in the rumen from escaping and she will die soon after.

Treatment

Treatment consists of a subcutaneous injection of calcium borogluconate, usually by means of a flutter valve.

There are several types of flutter valve. The most common type consists of a rubber tube (see Fig. 19.1) that attaches to a large-bore needle at one end and attaches to the bottle of calcium at the other. As the liquid flows from the bottle down the tube and into the cow, a small rubber valve allows air to leak into the bottle, so avoiding a vacuum and maintaining the flow of the liquid.

There is a number of calcium borogluconate preparations available, some are 20 per cent solutions, others are 40 per cent solutions, while others have added magnesium and phosphorus. Your veterinary surgeon will identify the most suitable one for your herd.

In cases where the animal is close to death, the calcium borogluconate may be administered intravenously. This is a specialized technique which should be carried out by a veterinary surgeon.

Cows normally respond quickly to treatment and are up and walking within a few hours. The stockperson must ensure that the calf does not suckle too much milk from the cow, as this may induce another attack.

Prevention

1. Injections of vitamin D eight or ten days before calving help to mobilize calcium reserves from the bones (*see also* page 153).
2. Avoid a high intake of calcium during the dry period, as this upsets the calcium mobilization mechanisms.
3. Ensure that the cow is getting adequate magnesium and phosphorus, as an imbalance of these two minerals interferes with calcium uptake.
4. Feed extra calcium just before and just after calving.

Grass staggers (hypomagnesaemia)

A condition that affects both cattle and sheep. Again, as it is more common in cows than ewes, I will refer to cows, although the symptoms, treatment and prevention are similar for both species.

The disease is due to a shortage of magnesium in the blood which usually occurs as a result of low magnesium in the diet. Unlike calcium in the case of milk fever, the cow has no body reserves of magnesium that she can call on.

Symptoms

An affected animal is usually excitable and, in the early stages, exhibits a slightly stiff-legged walk. She will eventually fall over with her legs either 'paddling' violently or held stiff. She may foam at the mouth and have a wild-eyed look. If left untreated, the animal will die quite quickly.

Treatment

If grass staggers is suspected, help should be sought from a veterinary surgeon, as treatment from unqualified persons may kill the animal.

Treatment is usually by subcutaneous injection of magnesium sulphate.

Prevention

As magnesium is not stored in the body, the animal needs a regular daily intake, especially when grazing lush pasture in May, early June and September, which may be low in magnesium. Problems can also occur under harsh conditions when animals are wet and cold.

1. Adding calcined magnesite to the daily ration may help, but it is not very palatable and may depress the intake of food to which it has been added.

2. Magnesium lick blocks are available, but some cows never seem to go near them.
3. The pasture can be dusted with calcined magnesite, although this is time-consuming and labour-intensive.
4. Magnesium can be added to the drinking water. This is a very effective method, as high yielding cows that require more magnesium will drink more water.
5. Magnesium 'bullets' are available. These metallic cylinders are administered using a dosing gun. They lodge in the reticulum and slowly dissolve over a few months, releasing magnesium as they do so.
6. Maintaining the correct soil pH helps to improve the magnesium uptake of the sward.

Acetonaemia (ketosis or slow fever)

This condition affects cattle whose diet is short of energy. Body fats are broken down to make up for this shortage, which results in chemical ketones being released into the body. Large amounts of these ketones are toxic, leading to the condition acetonaemia. It is called acetonaemia because acetone is the most characteristic of the ketones released.

Symptoms

Acetonaemia is often seen in high yielding cows, as it is they who are most likely to be short of energy. It may occur at any time of the year, but two to three weeks after calving is a common time. There is a mild form which may last several weeks in which the cow slowly loses body condition as time passes. In the more severe form the cow appears dull and listless, she will have a dull coat and have lost body condition. Her breath will smell of acetone (pear-drops), which can also be detected in the urine and dung. She will have hard, dry dung which is dark and often covered in shiny mucus.

Treatment and prevention

The disease is not usually fatal, although it can cause considerable economic loss due to a fall in milk yield at the peak of lactation. Successful treatment and prevention relies on identifying why a shortage of energy exists.

Immediate treatment will normally consist of an injection or drench to boost blood sugar levels and an injection to raise the metabolic rate of the body. Glucose is only useful if it is injected directly into the blood stream. If administered by mouth, it is broken down in the rumen.

Long-term prevention must take into account the dietary needs of the animal. It should be noted that cows that are overfat at calving have an increased risk of acetonaemia due to the fatness causing a reduced appetite in early lactation and therefore a reduced energy intake.

Pregnancy toxaemia (twin lamb disease)

Pregnancy toxaemia in ewes has a similar origin to acetonaemia in cows. Ketosis is certainly present and the problem of carbohydrate metabolism is similar. The problem is brought about, not by producing high yields of milk as in cows, but by

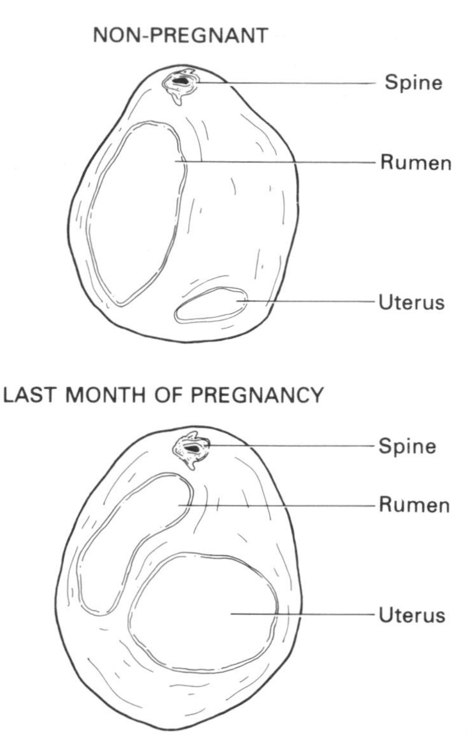

FIG. 19.2 Vertical section through the abdomen of the pregnant and non-pregnant ewe.

producing two or more fetuses. The disease occurs in the latter part of pregnancy, usually in the last month when the fetuses are growing at their fastest and their demands on the ewe are at their highest.

The cause of pregnancy toxaemia is normally slight malnutrition of the pregnant ewe. This malnutrition can be brought about in a number of ways.

The bad weather that occurs in January and February can be one predisposing factor. The ewes are feeding quite well soon after tupping (October time), but when better feeding is required towards the end of pregnancy, frost and snow cuts down on the natural fodder available. Also a heavily pregnant ewe, particularly with a uterus carrying two or more lambs, has less space in the rumen, as the enlarged uterus encroaches on the available space in the abdominal cavity (Fig. 19.2).

When the ewe is too fat the chances of pregnancy toxaemia are increased, as the ewe will limit her own intake of fodder.

Symptoms

One or more heavily in-lamb ewes separate themselves from the flock. They may appear blind and show no signs of moving even when approached. They may stagger about, stumbling into objects. Tremors occur at times and convulsions may follow. There may be the sweet smell of acetone on the breath and constipation is quite common.

After the ewe collapses, death occurs quite quickly, in a matter of a few hours or a day. In a small number of cases the animal may survive for up to a week.

Treatment and prevention

In the early stages of the disease, Caesarean removal of the lambs by a veterinary surgeon may save the ewe, but treatment of well-progressed cases is often difficult.

The main preventive measure consists of proper nutritional management which will vary from farm to farm depending on the environment and whether an intensive or extensive system of flock management is adopted.

Fatty liver syndrome (of cattle)

This is a disease of high-yielding dairy cows that may occur soon after calving. It is closely associated with acetonaemia, as it involves the storage of body fat.

When an animal calls upon its body reserves of fat to make up an energy deficit, some fat is stored in the liver until it is broken down and used. If too much fat accumulates in the liver, it will eventually affect the normal functioning of the liver and the cow will show signs of liver failure.

Symptoms

Affected cows look dull and very lethargic and may stop eating. Severely affected animals may be unable to get up and any treatment may be useless.

Other diseases, such as milk fever, may be a contributory factor to fatty liver. When she is recovering from milk fever her energy intake is lower and stored fat will be mobilized to meet the energy demand with the result that fatty liver may occur.

Prevention

It is now accepted in the dairy industry that many cows may be mildly affected by fatty liver syndrome leading to reduced fertility of the affected animals and a greater susceptibility to disease.

Prevention is mainly a case of good nutritional management with cows being fit but not fat in late pregnancy. This is achieved by a good feeding in early lactation to avoid a reduction in body condition and subsequent weight loss.

Mineral deficiencies

Some sixteen elements are known to be essential to the animal body. Magnesium deficiency has already been discussed (see Grass staggers, page 147). This section will deal with other minerals that may be deficient under certain circumstances.

Calcium deficiency

Calcium has already been mentioned under the section entitled Milk Fever (page 145). However, in cases of milk fever, calcium is not deficient, it is temporarily unavailable. True calcium deficiency is sometimes seen in growing animals as rickets.

Calcium is a major constituent of bone and teeth. Any deficiency results in malformation or poor growth of bone. The symptoms of rickets include poor growth rate, lameness, stiffness in the joints and bone fractures.

Adequate levels of vitamin D in the diet are important in maintaining correct blood calcium levels. Because vitamin D is manufactured in the animal's body by the action of sunlight on the skin, any animals kept in poorly lit buildings and fed an inadequate diet are prone to rickets.

Copper deficiency

Sheep are very prone to copper deficiency which results in a condition known as 'swayback'.

Swayback in sheep

Swayback affects new-born lambs, the name being derived from the characteristically uncoordinated gait, the back legs swaying from side to side.

It is caused by the gradual destruction of some brain cells whilst the lamb is in the womb. This in turn is associated with low levels of copper in the ewe, but not necessarily because the pasture is deficient in copper, as it has been shown that some breeds of sheep are more prone to swayback than others.

In mild cases nothing abnormal may be seen for the first few weeks of life, but when the lambs are running about an affected animal may suddenly lose control of its hind legs, keel over and lie on its back. It will soon get up again, but it may have a strange walk for the rest of its life.

Treatment is of no use, but mildly affected animals may be able to be fattened successfully. Copper is extremely toxic to sheep, so any preventive measures must be carefully controlled.

Prevention consists of giving a copper supplement to the ewe during the last three months of pregnancy. Slow-releasing boluses that lodge either in the reticulum or abomasum (depending on the product) give good results, whilst mineral licks and drenching are not so satisfactory.

Copper deficiency in cattle

Copper is involved in the formation of red blood cells, bone growth and the colouring of skin and hair.

Cattle suffering from copper deficiency are dull, unthrifty and in poor condition, often scouring. The coat takes on a reddish tinge, giving the animal a rusty look. The hair around the eyes may go grey, giving the animal the appearance of wearing 'spectacles'.

Copper deficiency can be a problem on pastures where the soil is rich in molybdenum and sulphur, as these affect the absorption of copper. These pastures are known as 'teart' pastures and occur, among other places, in the county of Somerset.

Dairy concentrates normally provide adequate copper for milking cows, although high-copper formulations are available.

Slow-release preparations that provide copper over a long period include injections, a bolus that lodges in the reticulum and another type that lodges in the abomasum.

Pastures and newly ploughed fields can be top dressed with copper salts.

Copper and Pigs

Pigs are more tolerant of copper in their bodies than are cattle and sheep. Various pig-fattening rations contain small amounts of added copper that act as a growth promoter. It is important that cattle and particularly sheep do not gain access to such foods.

Cobalt deficiency

Cobalt deficiency seems to affect sheep more than cattle. It is used by the microbes in the rumen to synthesize vitamin B12. Cobalt-deficient areas are well known and are associated with the red sandstone and granite areas of Cornwall, Devon and Derbyshire. Symptoms include anaemia, poor growth rates and an increased incidence of infections. Prevention is by the use of long-acting boluses and an increase of cobalt in the diet.

Iron deficiency

Iron is an essential constituent of red blood cells. Most animals obtain an adequate supply of iron in their normal diet. However, young piglets have a low reserve of iron in their livers and are fed solely on milk that is low in iron, so consequently may develop anaemia. Symptoms begin to show at about fourteen to eighteen days of age, with affected animals being pale, almost white skinned, generally unthrifty with a poor growth rate. If the pigs are weaned at twenty-one to twenty-eight days and put on solid food, the symptoms will disappear as the food has added iron. However, the damage to growth rates has been done and the animal will always be 'behind'.

Prevention can be carried out in a number of ways and is normally done between birth and three days of age. An intramuscular injection of an iron preparation is popular, as are specialist drink supplements that not only provide iron but also encourage certain bacteria to thrive in the intestines.

Piglets reared out in the field do not suffer from anaemia, as they will pick up sufficient iron from their rooting in the soil. Indeed, an old cure for piglets suffering from anaemia is to throw a clod of earth into the pen and let the piglets play with it.

Phosphorus deficiency

Combined with calcium, phosphorus is a major constituent of bone. Phosphorus deficiency can occur on certain types of pasture and in dairy cows producing 15 litres of milk or more per day living on forage alone. The symptoms may show up as rickets, together with poor growth rate and weight loss.

Opinion is divided in the dairy industry as to whether phosphorus deficiency causes temporary infertility in cows. It seems that the calcium-to-phosphorus ratio in the diet is an important factor in maintaining herd fertility and, as a guide, the figure usually quoted is a ratio of 1:1.

Zinc deficiency

This deficiency is more common in pigs than cattle and sheep and results in the skin disease parakeratosis.

It begins with red pimples appearing on the skin of the flanks and abdomen. The skin gradually becomes thickened with thin yellowish or greyish scales.

Treatment is by means of a zinc cream applied to the skin and a correction of zinc levels in the diet.

Zinc is also associated with growth of the hoof horn in cattle and sheep. Supplements of zinc to the diet have been shown to improve the condition of the feet.

Vitamin deficiencies

As with minerals, vitamins are a vital part of an animal's diet, even though the actual quantities required of these organic compounds are minute.

They are divided into fat-soluble vitamins (A, D, E, and K) and water-soluble vitamins (C and the B complex).

This distinction has a practical bearing both on the source of the vitamins and because of the fact that the fat-soluble vitamins are the ones of greater importance to farm animals.

Vitamin A

A deficiency in vitamin A results in stunted growth and poor resistance to disease, especially those that affect the respiratory and digestive systems. Night blindness may occur, leading to irreversible total blindness if the deficiency is not corrected. All classes of farm animals are susceptible.

The main sources of vitamin A are oils of animal origin, for example fish liver oils. These are added to pig and poultry rations if required. Cattle are able to get their vitamin A from the yellow pigment carotene which is present in all green plants. However, carotene is easily lost by oxidation, so weathering and bleaching of grass and overheating of hay will all reduce levels within the plant.

Vitamin D

Vitamin D has a function in the absorption of calcium and phosphorus from the intestines and their deposition to form bone. Parathyroid hormone is also involved in this process.

Deficiency of vitamin D in the growing animal results in rickets, although an actual deficiency of calcium and phosphorus will also cause the disease.

The symptoms of rickets are swollen joints, bent and deformed limbs and easily broken bones.

Vitamin D is manufactured by the action of ultra-violet light on the animal's skin. A deficiency is possible if young growing animals are kept in poorly lit buildings and fed an inadequate diet.

Most pig and poultry rations have vitamin D added as a matter of course.

Vitamin E

Vitamin E is present in grass and other green foods. It remains in grass that is made into silage, but is completely lost from hay.

Vitamin E is involved with nerve and muscle function and a deficiency may result in cases of muscular dystrophy or 'white muscle disease'. This disease affects cattle and sheep. Symptoms include sudden death, stiff back legs as the animal walks, exaggerated breathing and, occasionally, blood in the urine.

Vitamin E has a complex relationship with the mineral selenium. It is known that, in areas where the soil is deficient in selenium, muscular dystrophy can be prevented if selenium is added to the ration.

Vitamin K

Vitamin K deficiency is most unlikely, as it is present in many foods and can also be synthesized by microbes in the rumen.

Vitamin K is involved in blood clotting. Problems can occur in cases where animals have been poisoned by warfarin rat poison or mouldy clover hay, as these contain a group of chemicals known as dicoumarols which prevent the action of vitamin K, so the blood will not clot. Symptoms of this include cuts which will not stop bleeding and large areas of haemorrhaging around the mouth and nose.

The vitamin B complex

Originally it was thought that there was just one vitamin B, but research has shown that many forms exist. Thus they are referred to as the vitamin B complex.

Cattle and sheep are not usually susceptible to vitamin B complex deficiencies as the vitamins are able to be synthesized by the microbes in the rumen. However, recent research has suggested that adequate levels of nicotinic acid may not be synthesized and that a supplement to the diet may be of benefit.

At one time a deficiency of nicotinic acid was thought to be responsible for 'greasy pig disease', a skin disease usually affecting piglets and weaners, but this is now recognized as a bacterial infection (*see* page 140).

Biotin deficiency in pigs can give rise to foot problems in the form of cracks in the sole or wall of the hooves.

Vitamin B12 is normally added to pig and poultry rations as a matter of course. Ruminants whose diet may lack cobalt have B12 added to their ration, as B12 and the mineral cobalt are closely linked.

Vitamin C

Farm animals are capable of synthesizing their own supply of vitamin C. Thus deficiencies are never seen.

Metabolic profiles

Metabolic profile tests have been developed for dairy herds. They are used if herds are found to be suffering from 'metabolic' types of problems. An indication of this would be the herd having more than the average number of cases of metabolic diseases, mineral deficiency diseases, poor fertility or poor milk quality.

The test involves taking blood samples from a representative group of animals and examining the samples for levels of the various metabolites in the blood. These include: glucose, albumin, globulin, urea, phosphate, calcium, copper, magnesium, sodium, potassium and haemoglobin.

The results are then interpreted by comparison with 'normal' values. Modern automatic analytical equipment that is computer-linked allows quick, large-scale analysis of data.

These tests must be carefully planned and carried out under the supervision of a veterinary surgeon, as a random do-it-yourself approach may not provide any useful information.

Digestive disorders

Acidosis

Acidosis affects cattle that are fed on forage and concentrates. The normal pH of the rumen is 6.0 to 6.5. If the diet is high in concentrates, the bacteria in the rumen flourish on the fermentable carbohydrates contained in the concentrates. Some of these bacteria produce lactic acid, so reducing the pH of the rumen. When the pH reaches about 5.0 the contracting and churning action slows and eventually stops. The cow will stop eating and begin to look as though she is in some discomfort.

In extreme cases, if the pH falls to a very low level, body fluids are drawn out of the circulatory system by means of osmosis and the cow begins to go into a state of shock, her blood pressure will fall and she will be panting.

Normally the cow produces vast quantities of saliva that contain bicarbonate. This prevents the pH of the rumen from getting too low.

There are three main ways that the stockperson can help to prevent acidosis:

Feed concentrates little and often.
Add sodium bicarbonate to the ration to counteract the lactic acid.
Ensure that the cows have access to some form of long fibre, e.g. hay or straw, as this slows down the fermentation process in the rumen.

Chronic acidosis can affect the butterfat content of the milk, as fewer acetic acid molecules (the basis of butterfat) are formed in the rumen as the rumen becomes more acidic. In addition, absorption of the toxic products of rumen fermentation can lead to inflammation of the laminae (*see* page 112), resulting in laminitis.

Bloat

The normal process of ruminal digestion (fermentation) results in the continuous production of gas that needs to be released by belching.

If the gas cannot escape from the rumen (which is situated on the left side of the animal), it will begin to swell. As the condition gets worse, swelling can be seen on both sides, but it is more pronounced on the left. The animal is in obvious pain and has a stiff-legged stance. Eventually the animal will fall down and may die from either heart failure or suffocation. The latter occurs because the contents of the rumen may be forced back up the oesophagus and inhaled into the trachea.

Bloat affects both cattle and sheep, but is more commonplace in cattle. However, sheep with heavy fleeces are prone to rolling over on their backs and being unable to get up. Gases in the rumen are unable to escape and the animal will 'blow'. Once the sheep is put back on her feet she is normally none the worse for the experience.

Bloat can occur in three ways:

An obstruction in the oesophagus (this does not normally affect sheep).
The rumen will not contract and so gas cannot be released.
Froth (containing gas) may build up in the rumen.

Obstruction

This is usually caused by something like an unchewed potato that gets lodged in the oesophagus. It is normally removed by the use of a specialist instrument known as a probang. This consists of a stiffened nylon tube that has a metal ball on the end. It is inserted into the oesophagus to force the obstruction down into the rumen.

No ruminal contractions

This appears most commonly in the weaned calf and is caused by milk getting into the under-developed rumen where it ferments. Gas is produced but cannot be released, as the rumen is unable to contract.

Frothy bloat

Some pastures, particularly clovers and lush leys, are prone to causing this type of bloat. Instead of gas collecting in the upper part of the rumen and being released, it collects in small bubbles that will not burst. As more bubbles are formed, so the bloat gets worse, even though the rumen is contracting normally.

The treatment of bloat

The object of any bloat treatment is to relieve the pressure of gas in the rumen. As the type of bloat causing the problem is normally not known to start with, a bloat drench is usually administered first. If a proprietory bloat drench is not to hand, 500 ml of linseed oil will do as a substitute, the idea being to break the surface tension of any bubbles holding the gas. If this does not work, then a pipe passed down the oesophagus that bypasses the rumen sphincter muscle may be effective.

If these actions fail and the animal goes down, the gas must be released using an instrument called a trocar and cannula (see Fig. 19.3). The trocar fits inside the cannula and, when together, they are stabbed firmly into the left flank of the animal 5 cm behind the last rib and 15 cm down from the spine. The cannula is left in place whilst the trocar is withdrawn, thus leaving an airway for the gas to escape.

This latter course of action is for emergency treatment only, as the wound can quickly become infected and lead to peritonitis which could be fatal.

If frothy bloat is the cause of the problem then the stomach tube or the trocar and

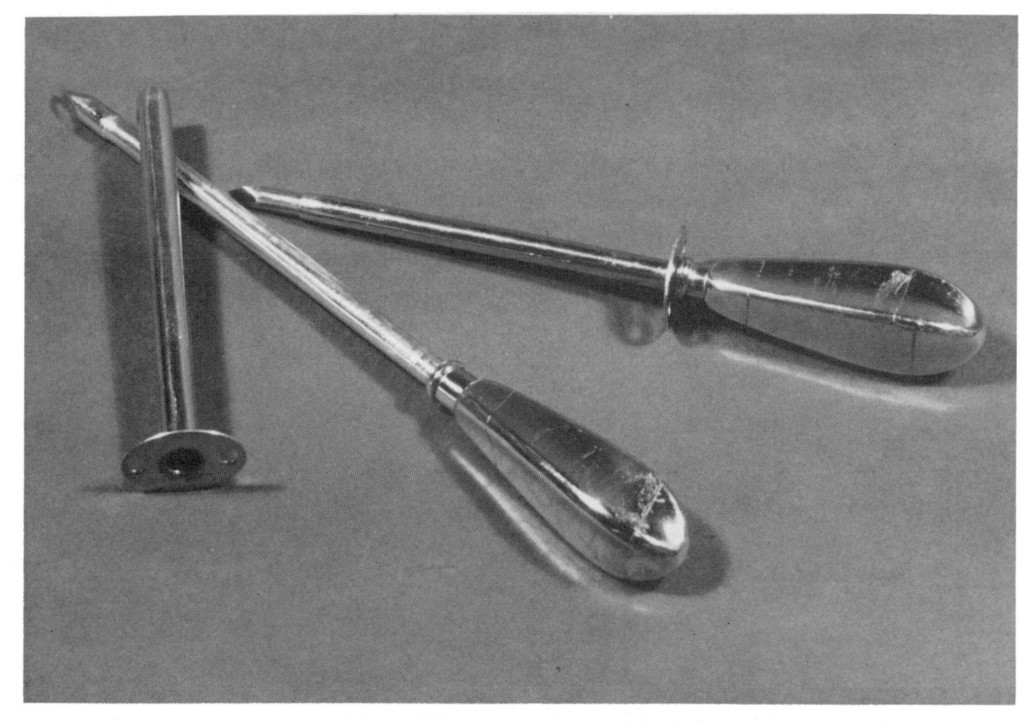

FIG. 19.3 A separated trocar and cannula and a cannula with the trocar inserted in it. They are used in the treatment of bloat.

cannula methods will not work. Thus the first course of action should be to administer a bloat drench.

Prevention

If pastures are known to cause bloat, they may be sprayed with mineral oils. The chemical paloxalone can be added to the drinking water. This will stop froth forming in the rumen.

Chapter 20

The clostridial diseases

The clostridia are a group of spore-forming bacteria that thrive in anaerobic conditions, that is they can multiply in the absence of oxygen. The spores are very resistant to many forms of disinfection and can persist in the soil for many years.

Because clostridia can live without oxygen, they are able to infect deep or closed wounds within the body and attack internal organs, for example the liver.

Clostridial infections can occur in all mammals (including man), but sheep are the most common farm animals to be affected, followed by cattle and occasionally pigs.

The clostridial diseases of sheep

There are seven main clostridial diseases of sheep. These are shown in Table 20.1, together with the relevant species of *Clostridium* causal organism.

As the diseases are linked by similar causal organisms, I will deal with prevention as one topic before mentioning the specific diseases.

Prevention

The main preventive measure available to farmers is vaccination. There are a number of products on the market that protect sheep against all the important clostridial diseases. These vaccines are formulated so that injection of just one substance will protect against the common diseases, and are the so-called seven-in-one or eight-in-one vaccines. (The seven-in-one vaccine protects against the clostridial

TABLE 20.1 *Clostridial diseases of sheep.*

Disease	*Clostridium* species
Pulpy kidney	*Cl. perfringens* type D
Lamb dysentery	*Cl. perfringens* type B
Struck	*Cl. perfringens* type C
Tetanus	*Cl. tetani*
Braxy	*Cl. septicum*
Black disease	*Cl. oedamatiens* B
Blackleg	*Cl. chauvoei* or *Cl. septicum*

diseases whilst the eight-in-one protects against the clostridial diseases and pasteurellosis—an infection of the lungs.)

To give full immunity to a flock a vaccination programme is drawn up in consultation with a veterinary surgeon. The timing of the vaccination is important because a number of the clostridial diseases affect lambs and it is essential that new-born lambs receive adequate levels of colostrum with the associated antibodies that it provides. It is now accepted practice to inject pregnant cows with a clostridial vaccine before calving. When the cow calves the colostrum (containing antibodies) is collected and deep-frozen. The farmer has then got a supply of colostrum that contains antibodies against the clostridial diseases which he or she is able to feed (by bottle or stomach tube) to weakly lambs or lambs from multiple births that may not get colostrum from their mother.

Pulpy kidney

Pulpy kidney is an acute, fatal disease most common in lambs of between three and twelve weeks of age. However, it can also occur in newborn lambs and in sheep between six and twelve months of age.

It is caused by *Clostridium perfringens* type D which multiplies rapidly in the intestines, producing a powerful toxin which is absorbed into the blood stream causing paralysis of some major internal organs. Death occurs in a matter of hours. Because of its action in the intestines, the disease is also called enterotoxaemia.

The causal organism is present in the gut of healthy sheep. It is only when it is stimulated to multiply that problems occur.

The disease affects animals that are 'doing well', often strong, single lambs, but it can also cause losses when lambs are suddenly put on lush grass after a period of poor nutrition.

The name pulpy kidney comes from the fact that when the animal is examined post-mortem the kidneys appear to have degenerated to a pulp, although a number of other conditions can cause this symptom.

Prevention is by vaccination with a seven-in-one (or eight-in-one) vaccine.

Lamb dysentery

Lamb dysentery is also an acute, fatal illness that most often occurs in lambs under fourteen days old. The causal organism is *Cl. perfringens* type B, which occurs naturally in the bowel of lambs. The organism produces a toxin which is absorbed into the blood stream and can cause death in a very few hours. There are often no symptoms to be seen, although severe, sometimes blood-stained, diarrhoea, can occur for twenty-four hours or so before death.

Once a farm has been infected with the disease it will persist in the ground, buildings and pens for many years and, if it appears at lambing time, the number of affected lambs increases as lambing time progresses. This is because infective organisms are excreted and picked up by the new-born animals, especially in crowded conditions.

Prevention is by use of a vaccine, but an outbreak can be controlled by using a serum containing antibodies that react with the toxin.

Struck

Struck is very similar to lamb dysentery and pulpy kidney, as it is caused by a similar bacterium, *Cl. perfringens* type C. However, struck affects adult sheep in certain localities—mainly Romney Marsh and areas of North Wales—and seems to be worse when there is an abundance of food.

As with the two related diseases, struck kills quickly, there are often no symptoms, and prevention is by vaccination.

Tetanus

Tetanus ('lockjaw') affects sheep and lambs and is caused by the bacterium *Clostridium tetani*. It is a widespread bacterium found in the soil and in the faeces of animals, including horses and cattle.

Sheep are usually infected with the bacterium at birth, via the naval or via the wounds caused by tail docking and castrating. It seems that the use of rubber rings exacerbates the problem.

Symptoms appear a week or so after the wound has been made. At first the animal walks with a stiff-legged gait, then it will fall over, throw its head back and go rigid in a muscular spasm, then relax only to go into spasm again. Occasionally an animal will only reach the stiff-legged stage and recover, but if it progresses to the muscular spasm stage it is normally fatal.

Prevention is by use of the multi-vaccine as discussed earlier. Cleanliness of the equipment used for tail docking and castrating is also most important, as is general cleanliness and hygiene at lambing time.

Braxy

Braxy is an acute disease that affects sheep aged four to eight months, usually during the autumn months. It is more common in hill flocks than in lowland flocks. The causal organism is *Cl. septicum*.

It seems that one of the predisposing factors of this disease is the ingestion of frosted grass, as the organism multiplies in the walls of the abomasum. The stomach will then break down and the bacterium will enter the abdominal cavity.

There are often no symptoms to be seen and diagnosis is by post-mortem analysis.

It should be noted that if laboratory diagnosis is necessary the animal must be presented to the laboratory as soon after death as possible or other bacteria will multiply in the stomach and abdomen, making the analysis very difficult indeed.

Prevention is by vaccination.

Black disease

Black disease can affect sheep of any age, but normally adult sheep are most at risk. The causal organism is *Cl. oedematiens* B which inhabits the soil and healthy sheep.

The organism causes no problems until the sheep is invaded by liver fluke (*see* page 183). The damage caused to the liver tissue by the fluke seems to provide an excellent medium for the multiplication of the bacterium.

There are usually no symptoms, but if the animal is found dead with the front legs stretched out, black disease should be suspected.

The name 'black disease' comes from the fact that the inside of the skin surface turns black soon after death.

Diagnosis is by post-mortem analysis.

As black disease only occurs when liver fluke are present, it is usually confined to autumn, winter and spring and to lowland sheep grazing pastures infected with liver fluke.

Prevention is by control of the fluke and vaccination.

Blackleg or blackquarter (gas gangrene)

Blackleg is caused by *Cl. chauvoei*, whilst gas gangrene is caused by *Cl. septicum*. They are acute diseases which kill very quickly due to the powerful toxins the bacteria produce. Infection is normally via a wound, possibly from castrating or tail docking, but it can be caused by a dirty vaccination needle.

The name 'blackquarter' comes from the fact that the muscles of either the fore limb or hind limb swell as the bacteria multiply in these sites.

There are often no symptoms, as the disease kills very quickly, so post-mortem examination is used for diagnosis.

Prevention is by vaccination and the use of clean castrating, tail docking, ear tagging and vaccination equipment.

The clostridial diseases of cattle

Cattle are not affected by as many clostridial diseases as are sheep. However, when they are affected the illness is just as serious because the causal organisms are the same.

In cattle there are three main clostridial diseases:

Tetanus.
Black disease.
Blackleg (blackquarter).

Prevention

The main preventive measure is a vaccination programme for one or all three diseases. A single vaccine to protect against all three diseases is available.

Some farms are prone to tetanus and blackleg owing to previous cases of the disease on the farm and the ability of the bacteria to form spores that remain in the soil for many years. For such farms a vaccination programme is essential.

As the organism is in the soil and can enter the body via cuts and wounds, the stockperson should ensure that all wounds are cleaned and dressed and any operations (castrating, injections) are performed hygienically.

Tetanus

As with tetanus in sheep, the bacteria produce toxins which affect the nervous system. The first symptom is usually twitching of the muscles, which progresses to muscular spasms. The animal may not be able to swallow or move its mouth (lock jaw). Eventually the spasms increase until the animal goes down with its four legs and neck held rigid. Death follows quite quickly.

Blackleg (Blackquarter)

This is an acute disease which kills very quickly. If any symptoms are seen, the animal will be dull and will walk with a stiff leg. The affected area will be swollen and, if pressed, will produce a crackling sound like tissue paper, due to the gas produced under the skin.

After the animal dies, the affected area may burst open, releasing a frothy liquid that smells like rancid butter. This liquid contains the spores of the bacteria which will contaminate the soil.

Black disease

This is not a very common disease in cattle. It has the same causal organism as in sheep, but in cattle the bacteria are taken in via the mouth, usually with some contaminated food. It affects the liver and causes death very quickly.

Clostridial diseases affecting pigs

The main clostridial disease affecting pigs is clostridial enteritis caused by *Cl. perfringens*. It affects recently born piglets, the symptoms being a watery, occasionally blood-stained, diarrhoea and sudden death.

The bacterium is picked up by the piglet from contaminated surroundings, for example the farrowing crate, and from infected sow faeces.

It can be treated using an antibiotic, but the piglets will dehydrate very quickly and the disease can sometimes cause disastrous losses. Preventive measures include improvements in hygiene and possibly vaccinating sows with a clostridial sheep vaccine, although this must only be done under veterinary supervision.

Chapter 21

Other diseases of importance

This chapter contains descriptions of a number of diseases that, although important, do not fit neatly under the headings of previous chapters.

I have classed them according to the main species—cattle, pigs or sheep—that is affected.

Cattle

Bovine viral diarrhoea (BVD)

BVD is a viral disease of adult cattle, the same virus causing mucosal disease in young animals. Depending on the immunity level of the animal, the symptoms will vary from mild diarrhoea and refusal to eat to massive blood-stained scouring and destruction of the gut wall.

The virus attacks the mucous membranes of the body, including the mouth, throat, nose, trachea and lungs and, as mentioned, the gut.

As it is a virus there is no treatment, although secondary bacterial infections of the damaged membranes can be treated.

When first infected, animals may run a high temperature and, if milking, the milk yield drops and in some cases they may abort. After infection a level of immunity is built up, but some animals remain as carriers, infecting heifers and new animals that join the herd.

Pregnant cows that are infected with BVD in the first six months of pregnancy have an added problem. The immune system of the fetus is not developed until it is six months old. If the mother picks up the virus in the first six months of pregnancy it will cross the placenta, but the fetus will not produce antibodies against it. If, after being born, the calf is then exposed to the disease it will not manufacture antibodies and so will suffer a bad attack of the disease. There is no vaccine available.

Some farms have a policy of ensuring that all heifers are exposed to the disease before they are served.

Infectious bovine rhinotracheitis (IBR)

IBR is a viral infection of cattle that affects mainly the respiratory system, although other problems do occur. The virus can affect the eyes and can cause abortions in pregnant cattle at any stage of the pregnancy.

The most usual symptoms are dull animals off their food, having a high temperature with a nasal discharge. Some animals are so severely affected that they die, others have very mild, sometimes unnoticeable, symptoms. There is no treatment, although there may be a secondary infection of bacterial pneumonia which can be dealt with by using a suitable antibiotic.

Prevention is by annual vaccination with an attenuated live IBR vaccine. This is administered to the animal's nose by a nasal spray attached to a syringe. The vaccine multiplies in the nose to give rise to the immunity.

The disease is so widespread that many farmers vaccinate as a routine. This is especially the case in areas with a high density of cattle.

New Forest disease

New Forest disease is a bacterial disease of the eye caused by the bacterium *Moraxella bovis*. It can affect all ages of stock, but animals of about one-year-old seem to be most at risk.

In mild cases, tears that contain antibodies are produced to wash away the bacterium. In more severe cases the whole of the eye may ulcerate, turn white and all vision is lost. An antibacterial cream applied to the eye works well, but needs to be applied over at least four days. Nowadays some veterinary surgeons inject a long-acting antibiotic behind the membrane surrounding the eye, the conjunctiva.

Johne's disease

Johne's disease affects both cattle and sheep and is caused by *Mycobacterium johnei*.

The disease is picked up by the calf soon after it is born, but because it has an incubation period of two years, the symptoms are not seen until the animal is at least two years old. The disease is then brought on by some form of stress, for example, calving, and results in thickening of the small and large intestines.

Because the absorption of nutrients and the reabsorption of water is much less efficient because of this thickening process, the cow develops a very watery diarrhoea associated with a deterioration in body condition and a huge loss in body weight. If left alone, the animal will eventually die, but it should not be allowed to reach this stage, casualty slaughter being preferable to a lingering death.

The disease is transmitted to young animals, often whilst they are still with their mother, via the mycobacterium present in faeces and in feed and water contaminated with faeces.

Some cows can become carriers of the disease. These may be the ones with poor growth rates and low levels of production that occasionally have bouts of scouring. These animals will be excreting the infection, causing the disease to infect others in the herd.

Prevention is to identify and cull infected animals and to avoid faecal contamination of the calf, its food and its water.

A vaccine is available, but it requires a special licence from the Ministry of Agriculture, because *Mycobacterium johnei* is related to the bacterium that causes tuberculosis (*see* page 80) and the vaccine may affect the results of the tuberculosis test carried out on dairy herds.

Pigs

Bowel oedema

An oedema is a fluid-filled swelling, in this case in the bowel. The bacterium *Escherichia coli* is present when the disease occurs, but the condition is initiated by some form of stress on the pig.

It is most often seen one or two days after piglets have been weaned or moved to different pens. The symptoms are difficult to identify, with sudden death often being the only sign. Uncoordination, a high-pitched squeal, diarrhoea and swelling of the eyelids may also occur.

A change in the management of weaned or relocated pigs rather than the use of antibiotics will often prevent future cases. Providing straw bedding will help to reduce the incidence of the disease, as the increased intake of fibre helps the intestines to function effectively.

Swine erysipelas

Swine erysipelas is caused by the bacterium *Erysipelothrix rhusiopathiae*. There are two forms of the disease—acute and chronic. In both cases the bacterium gains entry to the body via cuts and scratches in the skin.

Acute erysipelas produces the classic symptoms of diamond-shaped blotches on the skin over the back accompanied by a very high temperature and an unwillingness to stand. In a few cases it can lead to sudden death. Pregnant sows may abort due to the very high body temperature.

Chronic erysipelas is characterized by stiffness in the joints and cold, purple ears. The stiffness is caused by the bacteria localizing in the joints, whilst the purple ears are caused by growths on the heart valves which leads to poor circulation of the blood.

Treatment is by administering antibiotics under veterinary supervision. There is a very effective vaccine available which is administered to all breeding stock (i.e. sows and boars) twice a year.

Growing pigs and bacon pigs are susceptible to the disease, but it is most often seen amongst breeding stock.

Often the disease will flare up (in unvaccinated pigs) after some form of stress—possibly mixing sows together or even a spell of very hot weather.

Sheep

Footrot

For the purposes of this section the term footrot refers to an infectious disease of the sheep's foot caused by two bacteria—*Bacteroides nodosus* and *Fusiformis necrophorus*. It affects sheep of all ages, although older animals have more serious symptoms than lambs.

The disease starts on the interdigital skin (i.e. between the two digits of the foot), leaving a raw area on both digits. It is quite sore and some animals can be seen grazing on their knees. If the disease is allowed to continue, the bacteria will invade the hoof area causing separation of the wall from the coronet.

When the feet are closely inspected they will look overgrown with foul-smelling decay under the excess growth. At this stage the sheep is severely lame and has difficulty walking around.

In order to control the disease effectively the stockperson needs to have some knowledge of the two bacteria involved.

Bacteroides nodosus and *Fusiformis necrophorus* are synergistic, that is they work together. If one or other of the two is not present, the disease does not have such a dramatic effect on the foot. *F. necrophorus* is a very common bacterium with many hosts, so it is difficult to eradicate. On the other hand, *B. nodosus* is only found in sheep (and goats) and will not live for more than two weeks away from the animal's body. Both bacteria are anaerobic, so exposure to air will kill them and both are sensitive to antibiotics and other chemicals. So by careful control of *B. nodosus* many problems can be solved.

Careful examination of the sheep's feet, coupled with paring away any separated horn, will help to treat lame sheep. This exposes the decaying area to the air which aids healing. It can also be sprayed with an antibiotic spray. Foot bathing using antibiotic preparations or zinc sulphate will help to kill bacteria and keep the feet in good condition. Some farmers use formalin in the foot bath, but this is very harsh and, although it kills bacteria, it does not promote the healing of tender skin.

Once out of the foot bath the sheep should, if possible, be put on to pasture that has been free of sheep for two or more weeks.

A vaccine is available that will protect the sheep against *B. nodosus*. As the timing of administration is important, the vaccine should be given under the supervision of a veterinary surgeon.

The incidence of footrot is influenced by a number of factors:

Wet fields or wet bedding aids the spread of the bacteria.
Warm weather encourages the bacteria to multiply.
A high stocking density gives more sheep the opportunity of picking up the bacteria.

It is possible to eradicate the disease by regular inspection of all feet and the separation and treatment of those affected. After undertaking this time-consuming procedure, the stockperson must ensure that the disease is not re-introduced by new stock, or by other sheep straying onto the fields.

Orf (contagious pustular dermatitis)

Orf is a virus disease of sheep that affects older lambs, although ewes and their suckling lambs can be affected. It causes lesions and scabs mainly on the lips and mouth, but it can spread to the inside of the mouth, the tongue, the nose and the legs.

Once the virus has caused a break in the skin, the bacterium *Fusiformis necrophorus* (*see* Footrot) enters, causing secondary infection.

If young lambs have scabs around the mouth, their lips can be very painful, so they will not suckle properly, causing them to lose body condition. This can lead to the ewe's udder and teats being affected, eventually causing mastitis.

Orf is spread by contact, but how the disease persists on a farm from one year to the next is not clearly understood.

Treatment consists of spraying or bathing the affected area with antibiotics to relieve the secondary infections. The virus will clear up of its own accord in three to four weeks.

A live vaccine is available that is administered to the ewe, but as it does not give lambs complete immunity via the colostrum, its effectiveness has been variable.

Stockpeople should note that orf can infect humans, causing painful sores to the hands or face that can last up to six weeks.

Scrapie

Scrapie is a chronic, progressive disease of sheep (and goats) that affects the central nervous system, especially the brain. At one time it was thought to have been caused by a virus, but research has indicated that the cause may be a protein-like structure. Whatever the agent is, it can withstand heat, disinfectants and ultra-violet light.

Scrapie has been present in Europe for more than 200 years, but it is not normally seen in Australia or New Zealand—the only cases reported being from imported European stock.

The disease has a long incubation period, up to three years, which has made identification and research very difficult. The first symptom that can be seen is restlessness, the affected animal rushing from place to place. This progresses to the most obvious signs which are scratching, rubbing and nibbling of the fleece. The rubbing leads to the fleece being ragged and broken with bare patches of skin, often with scabs.

Eventually the sheep will become uncoordinated, excitable and difficult to catch and examine. The symptoms may last weeks or even months, with the animal losing body condition all the time. The disease is always fatal so the stockperson should cull affected animals as soon as the disease is recognized. Diagnosis is by watching for the clinical symptoms, whilst examination of the brain tissue is needed for confirmation. There is no treatment available.

It is thought that scrapie can be transmitted from one sheep to another by contact and from mother to offspring in the womb, with a genetic factor in the infected sheep determining whether the disease develops or not.

Many pedigree breeding flocks have eradicated the disease by examining the family line of sheep with the disease and culling that line. Actual figures are hard to come by as any breeder admitting to having scrapie in the flock may suffer a reduction in sales of breeding stock.

Chapter 22

External parasites and their associated diseases

The dictionary definition of a parasite is 'an animal or plant which lives in or upon another organism (its host) and draws its nourishment directly from it'. Whilst this is accurate, it does not describe the damage, the irritation or the diseases that parasites can inflict on their animal hosts or the economic loss to the farmer.

This chapter deals with insect parasites (six legs, e.g. lice), acaridan parasites (eight legs, e.g. ticks) and one fungal parasite. Also described are the diseases that may be spread by the various parasites.

Insect parasites

Warble fly

This is a notifiable disease and so is dealt with in Chapter 11, page 83.

Lice

Lice are wingless insects 2–3 mm in length (Fig. 22.1) that feed by sucking blood from their hosts, by biting or by living on skin debris. Lice are highly specific to their particular hosts, so will only survive for a very short time on other animals—including man!

They affect cattle, sheep and pigs, causing intense irritation and general unthriftiness and making the animal more prone to picking up infections. A heavy infestation of blood-sucking lice can cause anaemia.

Lice infestations are heavier in winter than in summer. This is probably due to housed cattle and sheep spreading the lice between them, the coats being dirtier and the animals' nutrition not being as good as summer grass.

There are a number of treatments, including louse powder (benzine hexachloride, BHC), organophosphorus treatments such as that for warble fly, and ivermectin, usually used against worms but also effective against lice. Ivermectin is administered either by injection or by the 'pour-on' method.

None of the treatments is effective against the eggs that are attached to the hairs of

FIG. 22.1 The biting louse, *Haematopinus suis*, that affects pigs.

the host, so a knowledge of the life cycle is essential for effective treatment. In the case of cattle lice, the eggs hatch after about two weeks into immature lice which in turn take about two weeks to become mature lice. So, regular fortnightly treatments, or the use of a compound that will persist in the coat or body for more than four weeks, will give effective protection.

Keds

Keds are closely related to flies, although they are wingless. They are about 5 mm long, dark brown in colour and live on sheep (Fig. 22.2). They can be seen quite easily by parting the wool of infected animals. Keds suck blood and cause intense irritation to the sheep which will rub themselves on walls and fences, causing damage to the wool.

The life cycle normally takes place on one host, although the ked can live away from the sheep for about twelve days and the pupa may live on the ground for six weeks. One life cycle will normally be completed in a month.

Control is by dipping and shearing. It should be noted that keds may bite the person doing the shearing!

Blowfly

Blowfly attack all breeds of sheep normally during the summer months but very occasionally during winter housing. The condition is also known as 'blowfly strike'.

The blowfly or greenbottle lays its eggs on any decomposing moist material. This can be any wound on the sheep, but more usually it is the soiled area of wool, compacted with urine and faeces, on the breech and tail of the animal.

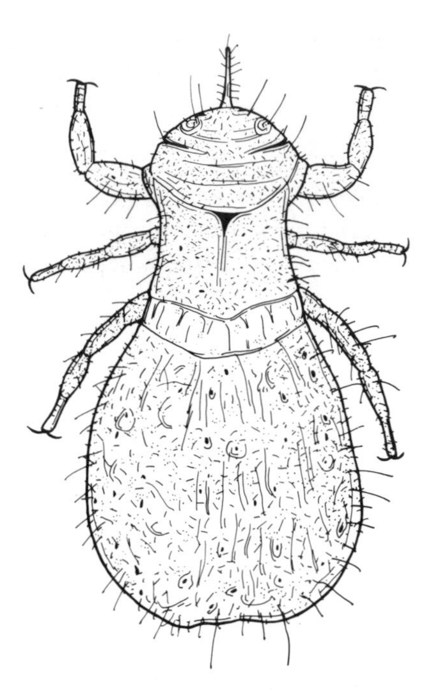

FIG. 22.2 The sheep ked, *Melophagus ovinus*.

The eggs hatch in twelve to twenty-four hours and the maggots are fully grown at about five days. The maggots, which are present in huge numbers, feed on the skin fluids, causing the tissue to break down.

The first signs of the strike are wriggling of the root of the tail and the sheep trying to kick or bite at the affected area. If the infestation is left untreated, secondary infection of the wound may occur.

Treatment is to clip the sheep in order to clean the area and to apply a chemical to kill the maggots.

Prevention is by the use of dips, especially when the weather is warm and humid, as this encourages large numbers of blowflies.

The routine job of 'dagging' sheep, that is removing contaminated wool from around the tail, is a valuable aid in the prevention of strike.

Headfly

The headfly, *Hydrotaea irritans*, is a biting fly that attacks sheep in an area around the English and Scottish border. It causes intense irritation to any open wounds on the head, especially around the base of the horns. It enlarges the wound and may introduce bacteria which cause a secondary infection, leading to illness and loss of condition.

The headfly has a four-stage life cycle of egg, larva, pupa and adult, with the pupa and adult active in the year following the laying of the eggs.

The flies are active around mid to late summer, especially in wooded and boggy areas.

In the past, the only means of prevention involved fitting protective hoods to keep the flies off, but now fly repellents are the main source of control.

The fly *Hydrotaea irritans* is also responsible for carrying bacteria to the teats of cows and heifers and may be involved in cases of summer mastitis (*see* page 127).

Acaridan parasites

Acarina are small animals which differ from insects in that the adults have eight legs, the body is not divided into segments and there is no marked difference between the larva and adult.

The most important acaridan parasites that affect animals are mites and ticks.

Mites

Mites are responsible for the skin diseases mange and scabies that can occur in all mammals and birds. The mite lives and feeds either on the surface of the skin or in the outer layers of the skin causing irritation, scabby skin and hair loss.

There are a number of different types of mite, but their life cycles are fairly similar.

The egg hatches in two to four days and produces a six-legged larva. After two to three days the larva becomes an eight-legged nymph which, three to four days later, becomes an adult (Fig. 22.3). The female is then fertilized so that in two to four days she will start laying eggs. The whole cycle takes seven to eleven days.

Mange in sheep

Psoroptic mange in sheep is the cause of the notifiable disease sheep scab and is dealt with in Chapter 11.

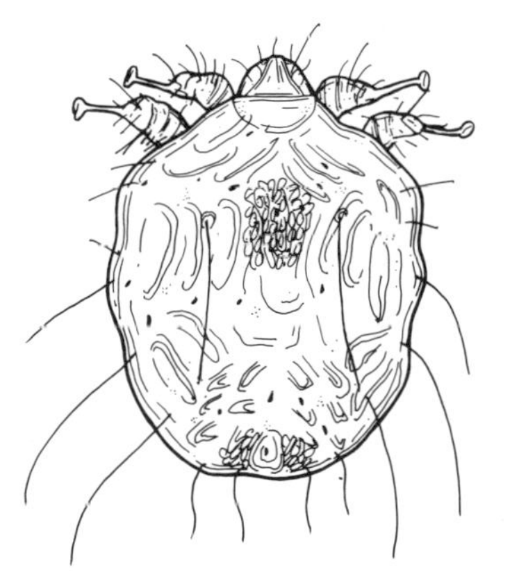

FIG. 22.3 Dorsal view of the female *Sarcoptes scabiei*, a mange mite.

Mange in cattle

Mange affects both adult cattle and calves, causing intense irritation and forming scabs and crusts on the surface of the skin, especially around the root of the tail and possibly over the neck.

One of the simplest methods of treatment is a 'pour-on' insecticide which is poured along the back of the animal and absorbed into the body.

It is normally effective against lice and even warble fly (*see* page 83), as well as mange.

Mange in pigs

Sarcoptic mange in pigs leads to yellowish scaly skin, constant scratching with subsequent skin irritation and, in severe cases, loss of body condition. It is often seen inside the ears, inside the front and back legs and on the shoulder blades.

It is carried by chronically affected sows and boars and is transmitted to piglets, usually affecting them after weaning. It can be brought into a herd by buying in infected stock and can also be present in buildings that have had pigs in recently.

Treatment is by means of washes, pour-on organophosphorus compounds, or by an injection of ivermectin.

It is possible to eradicate the parasite from a herd by means of a programme of ivermectin injections and by ensuring that only mange-free animals are purchased as replacement breeding stock.

Ticks

Although twenty-one species of ticks have been found in the United Kingdom, there are only two of major importance.

The first one is *Ixodes ricinus*, the so-called sheep tick (Fig. 22.4), although it will

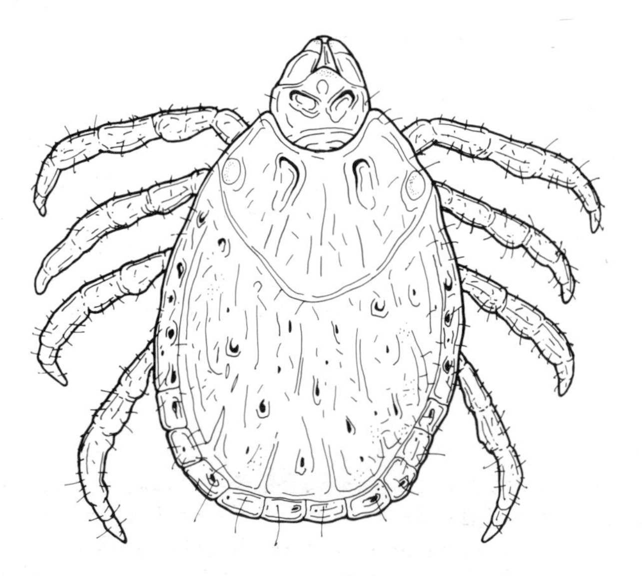

FIG. 22.4 The sheep tick, *Ixodes ricinus*.

Year 1	Year 2	Year 3	Year 4
Egg Laid in spring/summer on grass	Larva Attaches to animal in spring and then back onto grass	Nymph Attaches to animal in spring and then back onto grass	Adult Feeds on animal in spring,mates and lays eggs
overwinters →	overwinters →	overwinters	

FIG. 22.5 The life cycle of the sheep tick.

live on cattle, deer and even man. It is found in most areas of the United Kingdom, especially on upland and coarse pastures, including those that contain rushes and bracken.

The second, *Haemaphysalis punctata*, is only found in some coastal areas of South Wales, the Lleyn Peninsula and Anglesey.

The tick has a fairly complicated three-year life cycle as shown in Fig. 22.5. In its life cycle it will attach itself to three different hosts and may spend less than a total of three weeks in three years actually on the hosts.

An egg laid in some coarse vegetation in year one will hatch out as a larva in the spring of the second year. The larva will then climb to the top of some tall grass or bracken and wait until an animal brushes past.

It will attach itself to this animal, suck the blood of its host and, after about five days, drop to the ground where it remains for the rest of the year. In the third year the tick emerges as a nymph, repeats the feeding process and falls to the ground to overwinter. In the fourth year the tick emerges as an adult.

There are two distinct tick activity times. The first is in the spring and those that hatch then continue their lives as spring feeders. The second is in the autumn and these continue as autumn feeders.

Problems caused by ticks fall into two categories:

> The primary problem is caused by the tick itself. It causes irritation to the animal and in severe infestations could lead to anaemia due to loss of blood.
> The secondary problem is caused by the parasites that the tick carries. As the tick begins to feed on its host it passes a drop of saliva into the host. This saliva acts as an anticoagulant for the blood so that the tick can feed, but this action can also infect the host with a number of diseases commonly called tick-borne diseases.

Tick-borne diseases

Tick-borne fever

Tick-borne fever affects both cattle and sheep and is caused by a rickettsial organism that has some of the characteristics of both bacteria and viruses. It attacks and destroys some white blood cells, so reducing the animal's resistance to disease. In sheep the lowered body defences mean that they are more prone to other tick-borne diseases such as louping ill or tick pyaemia. Tick-borne fever may be responsible for some abortions in ewes, but the most usual symptoms are lethargy and fever which often pass unnoticed.

In cattle the symptoms are more pronounced. Affected animals may lose their appetite, have stiff joints and a high temperature. Milking cows will show a reduction in yield.

As with sheep, affected cattle are more prone to other diseases because of the loss of white blood cells.

Antibiotic treatment is usually effective for both cattle and sheep.

Louping ill

Louping ill is considered mainly as a disease of sheep, although it can affect cattle, pigs and deer.

It is a virus infection transmitted by the bite of an infected tick at either the nymph or adult stage.

The virus multiplies in the blood and then attacks the central nervous system. In the first stage of the disease the sheep may appear dull and have a high temperature. Many sheep will recover at this stage, showing no further symptoms. However, if the disease continues, the virus attacks the brain, resulting in symptoms of excitability such as uncoordinated muscular spasms, trembling and a high stepping walk. This is soon followed by coma and then death.

The disease is most common in lambs and yearlings because older sheep can gain immunity to the disease.

Sheep in well-known 'louping ill areas' can build up a strong immunity to the disease, so care must be taken when introducing louping ill-free stock into an infected flock.

There is no treatment for affected animals, but a vaccine is available as a preventive measure. If ewes are vaccinated a month before they are due to lamb, the lambs will acquire immunity via the colostrum.

The word louping is a Scottish term and means 'leaping', although actual jumping in the air is rarely seen.

It should be noted that shepherds, farmers and veterinary surgeons are all susceptible to infection.

Tick pyaemia

Tick pyaemia is caused by the bacterium *Staphylococcus aureas* which finds its way into the blood-stream of sheep following a tick bite. As the bacteria is already on the skin of the sheep, the tick is not the true cause of the disease.

The symptoms are very similar to joint-ill (*see* page 138), that is, lameness and uncoordination of movement leading to paralysis. Abscesses form throughout the body, especially in the area of the brain and spinal cord.

The only way of preventing the disease is to prevent the ticks attaching themselves to the fleece. This is done by the use of dips for the lambs, but as the persistence of the dip in the fleece is very short (seven to ten days) it is not a very satisfactory procedure.

Injections of long-acting antibiotic are useful for treating the odd case, but, if the disease becomes established, little can be done.

Redwater

Redwater is a disease of cattle caused by the protozoan parasite *Babesia divergens*, which is related to the parasite that causes malaria in humans. It invades the blood stream and begins to multiply in the red blood cells, causing them to rupture and release the red pigment haemoglobin. The red colour taints the animal's urine which can vary from slight red to almost black depending on the extent of the infection, hence the name redwater.

Because of the damage to the red blood cells, the animal may develop anaemia which, in severe cases, may require a blood transfusion.

There are specific drugs available to treat the animal, although a vaccine is not yet available in the United Kingdom.

The *Babesia* parasite is able to survive inside the tick from the egg stage to the adult stage and may pass onto the next generation of ticks via the egg, even though it may not have been anywhere near cattle or cattle's blood. Because of this, areas can remain infected for a long time.

Tick population control

It is possible to get an idea of the number of ticks in an area by dragging a light-coloured woollen blanket over the pasture during the spring or autumn tick season. The ticks will attach themselves to the underside, the number giving an estimation of the level of infestation.

Because ticks are multi-host parasites and can live on wild animals such as rabbits, voles and foxes, and because they are only attached to the host for a very short time period, they are extremely difficult to control. Sprays and dips work well but do not persist in the animal's coat for long. It is possible to dust or spray pastures, but because of the rough nature of the terrain and vegetation, it is expensive and not too successful. Improving the land by draining and ploughing is an effective method of tick control. However, this will lead to the flock having less immunity to tick-borne diseases such as louping ill which may then recur if the sheep are turned onto upland grazing at a later date.

Fungal Parasites

The only fungal parasite of major importance to domestic animals is ringworm.

Ringworm

Ringworm is a fungal infection of the surface of the skin which can be caused by a number of fungi, although typically it is caused by the fungus *Trichophyton verrucosum*. It can affect cattle, horses, dogs, cats, chickens and humans, especially young children. It is not usually seen on sheep or pigs.

Ringworm shows up as patches of dry, raised, crusty and hairless skin. The patches are often circular, but in severe cases the patches may join so that large irregular-shaped areas are formed.

In cattle it is very common among young animals that are housed during the winter. Their heads, necks and eye-lids are the areas that are most affected. Because ringworm is very irritating for the animal, it will rub and scratch itself which, in severe cases, may cause secondary bacterial infection, with pus and scabs forming.

The spores will live in dry conditions for three to four years, so batches of animals can become infected when they are moved into a contaminated building.

Painting a building with creosote or using a flame gun is effective as long as the job is done thoroughly.

There are two types of treatment for ringworm. The first is an in-feed medication, fed daily for seven days, that prevents the fungus growing on the skin surface. The second is a spray treatment that covers and soaks the whole of the animal. Whatever treatment is chosen, the whole group of animals should be treated as the incubation period is about three weeks. To try and treat individual animals is normally a false economy.

Animals in poor condition are often worst affected, so maintaining good health is essential. When young animals are turned out in the spring some cases of ringworm clear up without treatment. This is because the fungus is killed by ultra-violet light and with plenty of lush grass the animal is on a better plane of nutrition.

As mentioned earlier, ringworm can affect humans, so care should be taken when handling affected animals.

Chapter 23

Internal parasites

The object of this chapter is to consider the main disease-causing internal parasites of cattle, sheep and pigs.

The diseases caused by these parasitic worms can be divided into three main areas:

Parasitic disease of the lungs.
Parasitic disease of the digestive system.
Parasitic disease of the liver.

Major advances in the chemical composition of anthelmintics (i.e. chemicals effective against worms) have meant that the stockperson has a huge armoury of drugs available to kill parasitic worms at various stages of their development whilst still inside the animal's body. Nevertheless, many animals still carry a large worm burden, causing suffering to the animal and economic loss to the farmer.

In order to treat animals against these parasites, a basic understanding of the worm's life cycle is essential so that the treatment can be administered at a time when it will have the most effect.

Drugs are not the only weapon that can be used to fight worms however. Unlike bacteria or viruses, which reproduce and multiply inside the animal's body, female worms lay many thousands of eggs which are passed out in the faeces. No worms will become adults without first going outside the animal's body. This means that they must be picked up from infected faeces or whilst the animal is grazing. Therefore, general hygiene and grazing management have an important role to play in the control of parasitic worms and their associated diseases.

As with many diseases, healthy animals have a greater tolerance to worms than unhealthy ones. Adult stock may build up some immunity to parasitic diseases, so the worst cases are usually seen in young and growing stock.

Please note that, for the purposes of this book, the 'maggot' of the warble fly, which could be considered as an internal parasite, is dealt with in Chapter 11, Notifiable Diseases, page 83.

Parasitic disease of the lungs

Husk or lungworm

Cattle and sheep are both affected by this disease. In cattle the causal organism is the lungworm *Dictyocaulus viviparus*, in sheep it is caused by *Dictyocaulus filaria*. As the disease is more serious in cattle, this section will consider husk in relation to cattle.

The life cycle of the lungworm is fairly straightforward (Fig. 23.1). It lives in the bronchial tubes of the lungs where the female lays huge numbers of eggs. These eggs hatch into extremely small larvae (L1) which, because they irritate the animal's throat, get mixed with mucus and are coughed up into the throat/mouth. They are swallowed, pass through the intestines and out in the faeces. On the ground the L1 larvae develop into more mature larvae (L3), in about seven days if the weather is warm and humid. This development may take up to a month or more if the weather is cold.

Once mature the L3 larva uses the film of moisture on grass to gain access to the grass tip where it is eaten by the grazing animal, so finding its way into the intestines. The L3 larva then migrates to the lungs via the lymph and blood vessels. In the lungs the L3 larvae mature to adults, the female will lay eggs and the cycle starts again.

Symptoms

Lungworms mainly affect calves turned out to pasture for the first time, the calves having no resistance to the disease. Adults have usually built up some resistance, but a sudden heavy infestation may give rise to some cases amongst older animals.

For about two weeks after eating the larvae no symptoms are visible, but as they begin to attack the lungs the animal will pant and may have laboured breathing.

In severe cases, the calf will develop a deep throaty cough and will stand with an arched back, with its tongue protruding from its mouth. It may lose weight quickly and die. Strangely enough, an animal that is coughing may not be as severely infected as one that is struggling for breath. This is because the latter may be, quite literally, slowly suffocating due to huge numbers of worms blocking the respiratory passages.

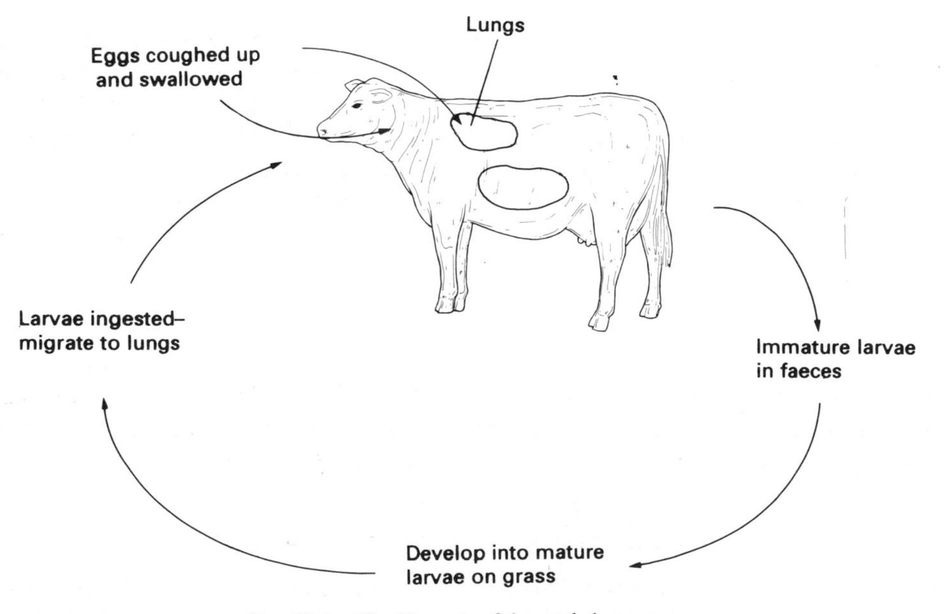

FIG. 23.1 The life cycle of the cattle lungworm.

Treatment

Affected calves must be moved onto a clean pasture or be brought indoors. They can be treated with an anthelmintic to kill the worms, but this is not without its problems. An injection of wormer may kill the lungworms, but these will remain in the lungs, possibly causing pneumonia.

Prevention and control

The best preventive measure is by vaccination. An oral vaccine is available (from a veterinary surgeon) that consists of irradiated lungworm larvae. The irradiation procedure prevents the larvae from migrating to the lungs so the animal will build up some immunity to the disease. It is a two-dose vaccine, the first dose given six weeks before the animal is turned out to grass and the second dose given two weeks before turn-out.

Animals need only be treated with this method in their first year of life. Subsequently, the exposure to lungworm will build up sufficient immunity.

It should be noted that other worm treatment programmes, such as injecting with ivermectin, can be used if desired, but should not be used together with the oral vaccine as the ivermectin will interfere with the vaccine, possibly reducing its immunity-building effect.

It is possible to build up immunity in young calves by having a strict programme of rotational grazing where, for the first year, calves are kept off grass that has been grazed by older stock. This will prevent the build-up of disease and means that calves will not suddenly take in large numbers of larvae. However, as the weather and not the stockperson often dictates the grazing programme, this procedure can be difficult.

Metastrongylus ***infection in pigs***

Adult worms in the lungs produce eggs that are coughed up, swallowed and passed out in the faeces. Here they hatch into larvae which then infect earthworms. An earthworm is then eaten by the pig and the infective larvae migrate through the gut wall to infect the lungs. Because intensively reared pigs do not have access to earthworms, the disease is normally confined to outdoor herds.

The symptoms are coughing that sounds like a dog's bark and poor growth rates. If pneumonia is present in the herd, the *Metastrongylus* worm will exaggerate the problem.

Ascaris ***infection in pigs***

Ascariasis is considered later in this chapter under the heading of 'Parasitic disease of the digestive system'. It should be noted however, that immature *Ascaris* worms migrate from the gut and can attack the lungs, causing pneumonia.

Parasitic disease of the digestive system

There are many different stomach and intestinal worms that infect cattle, sheep and pigs, but, because of modern anthelmintics and husbandry practices, only a few cause problems. This section will consider those parasites of the digestive system that are still of importance.

A disease or problem of the digestive system that is caused by parasites (for example, scouring) is known as parasitic gastroenteritis.

Ostertagiasis

The stomach worm *Ostertagia ostertagi* affects both cattle and sheep. The adult worm lives in the abomasum (fourth stomach) where it lays eggs that pass out in the faeces. These eggs develop, on the ground, into larvae (L3). These larvae are able to move to the top of grass blades where they are eaten by other grazing stock. The larvae then burrow into the gastric glands that line the wall of the abomasum, where they develop into adults. Once they reach maturity they begin to produce eggs. The cycle, from ingesting the L3 larvae to eggs appearing in the faeces, takes about three weeks.

The weather has a direct effect on the time it takes for the L3 larvae to mature when outside the animal's body. A low temperature delays the development of the larvae whilst warmer weather speeds it up.

This means that eggs passed out in the faeces in April or May can take five or six weeks to mature, whilst those passed out in June or July will mature in about a fortnight. The consequence of this is that around mid-July there are huge numbers of L3 larvae on the grass, resulting in heavy infestations of grazing stock.

Symptoms

The developing larvae damage the chemical-producing gastric glands of the stomach with the result that two of the chemicals required for protein digestion (hydrochloric acid and the enzyme pepsin) are not produced in sufficient quantities. The protein in the gut is not properly broken down or absorbed so the animal begins to scour. In calves the scour is a bright green colour, whilst lambs have heavy soiling of the wool around the tail. Very heavy infestations lead to weight loss, stunted growth and, if left untreated, may result in death.

Prevention and control

There are a number of factors that will influence the level of ostertagiasis in any year. These can be summarized as:

The level of 'carry-over' larvae on the pasture from the previous year.
The stocking rate—heavy stocking will result in more eggs dropping onto the ground and, ultimately, more larvae.
The season's weather—a cool summer will delay the larvae maturing. A hot, dry summer will kill the larvae on the grass.

Regular treatment of stock with anthelmintics at three-weekly intervals throughout the grazing season will kill the female worms before they produce eggs. However, as this can be a very labour-intensive operation, several animal health companies produce a 'bolus' wormer (see Fig. 23.2). This is a large tablet or pellet containing a slow-release anthelmintic. It is administered to the animal at the start of the grazing season, it lodges in the reticulum where it will release the wormer over a period of months.

The bolus treatment system is used to protect cattle and calves, but is not used for sheep and lambs.

Anthelmintic treatment should be used in conjunction with a clean grazing system. Calves and lambs should be turned out onto grazing that has not been contaminated by older stock. If animals are treated for a heavy ostertagia burden they must be moved to clean pasture or the larvae on the grass will be ingested and a new worm burden will be established. Older animals gradually develop an immunity to *Ostertagia*.

Winter ostertagiasis (Type II)

Ostertagia worms that affect animals in the summer are referred to as Type I, those causing problems in the winter are known as Type II.

Type II infection results from some *Ostertagia* worms over-wintering as larvae in the stomach lining. These are known as L4 larvae. As the L4 larvae emerge to become adults in April or May, there is a sudden rise in the animal's worm burden causing it to scour and lose weight.

Prevention is by dosing with anthelmintic when the animals are housed for the winter. Stock that are out-wintered should be treated in December.

Nematodiriasis

Nematodirus worms affect both cattle and sheep, but as they do not seem to cause much of a problem to cattle this section will deal with nematodiriasis in sheep.

Two species, *Nematodirus battus* and *Nematodirus filicollis*, cause the disease which affects lambs between six and ten weeks old, but can also affect older lambs.

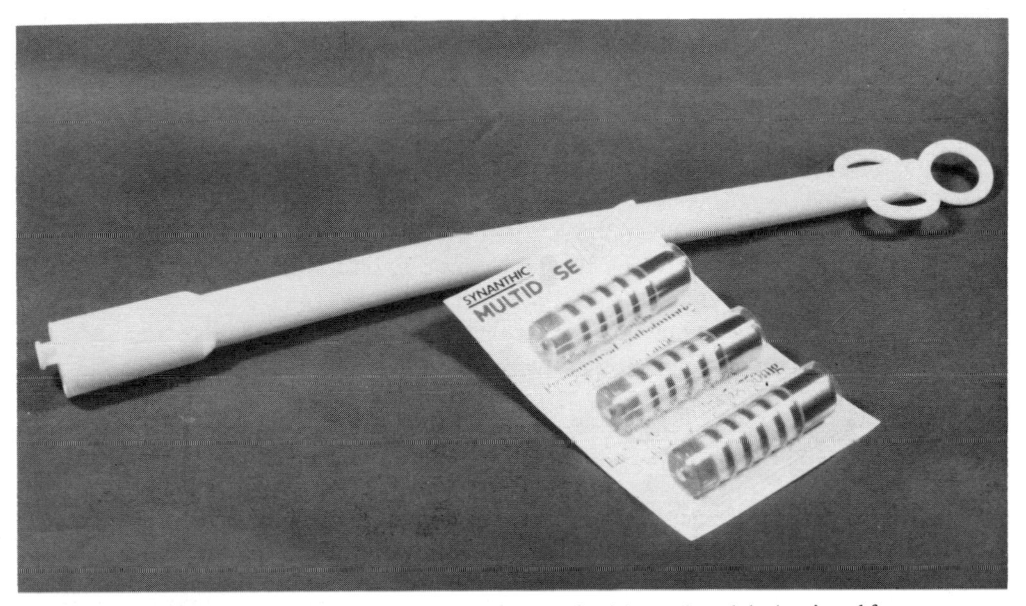

FIG. 23.2 A pack of three 'bolus'-type wormers for dairy cattle and the 'gun' used for administration.

N. battus eggs are passed out in the faeces over the summer, but, strangely, a temperature of below freezing followed by a temperature of 10 °C is needed before they will hatch. This means that large numbers of eggs usually hatch out in March, April and May. If there is a spell of early warm weather, say in February, the eggs will hatch but the larvae will die before the lambs have been turned out to grass. If, on the other hand, cold weather persists into the spring, the larvae will be active as the lambs begin to graze.

The eggs of *N. filicollis* are produced in April and will have hatched by June, so that infective larvae will be present on the pasture over the winter period.

The symptoms of nematodiriasis are sudden scouring, dehydration, little or no weight gain and, in severe cases, death.

Control is by the use of anthelmintics on a regular basis and by ensuring that new lambs do not graze on pastures used by lambs in the previous year. After infection the animals will gain immunity to the disease.

Each year the Ministry of Agriculture issues a nematodiriasis forecast based on soil temperatures. This should be interpreted in conjunction with the lambing dates on individual farms.

Stomach worms of pigs

There are three main stomach worms of pigs:

Ascaris suum, the large roundworm.
Oesophagostomum dentatum, the nodular worm.
Hyostrongylus rubidus, the red stomach worm.

Ascaris suum

The adults of this worm occur in the small intestine where they produce eggs that pass out in the faeces. These eggs can persist outside the pig for up to ten years. The eggs are ingested by the pig where they hatch, migrate to the liver and develop into third stage larvae. This stage of the life cycle gives rise to the typical 'milk spot' liver that is seen when the animal is slaughtered, i.e. white spots covering the surface of the liver.

The larvae then migrate to the lungs, become fourth stage (L4) larvae, and are coughed up, swallowed and return to the small intestine.

The disease causes economic loss by the fact that the livers may be condemned and it may cause scouring and lung problems in badly infected pigs.

Oesophagostomum dentatum

This worm is found in the large intestine of pigs. It has a relatively simple life cycle in that the adults produce eggs that are passed out in the faeces where they develop into infective larvae. These are ingested and burrow into the wall of the large intestine, forming a characteristic nodule.

A severe infestation may cause scouring, but normally reduced weight gains are the only symptom.

Hyostrongylus rubidus

This red worm mainly affects pigs kept outdoors as the development from eggs to larvae is dependent on the weather. Warm wet weather helps development, cold dry weather will kill the eggs.

The worm develops in the stomach lining and affects the production of the chemicals needed for digestion.

The disease is often seen in lactating sows where they continue to lose weight despite correct feeding. Affected animals do not normally scour.

Prevention and control

It is difficult to eradicate all stomach worms from a pig unit because, for example, the ascaris egg has such a long survival time. However, a regular anthelmintic treatment programme should result in a low worm burden.

Many pig farmers worm their sows about two weeks before they are due to farrow, then wash them as they are moved into the farrowing pens to remove any faeces that may contain worm eggs. This means that the piglets get a 'clean' start. Regular removal of dung and the pressure washing of pens and buildings will stop the build-up of eggs and infective larvae.

The worm burden of a herd can be monitored using egg counts from the faeces, information from the abattoir to which the pigs are sent and post-mortem examination of pigs that die.

Outdoor pig herds should have a land rotation policy to avoid the build-up of larvae on a particular area.

Parasitic disease of the liver

Liver fluke or fascioliasis

The liver fluke is a small flatworm, *Fasciola hepatica*, that affects a wide range of mammals (including man) but causes most problems to cattle and sheep.

The life cycle of the fluke is complex (Fig. 23.3). Adults in the liver produce eggs that pass out in the faeces. When the temperature is above 10°C and it is moist, the eggs hatch and release larvae called miracidia. The miracidia have the ability to swim until they come into contact with a small snail, *Lymnaea truncatula*. They penetrate the snail and once inside it multiply to form another type of larvae, cercariae. The cercariae leave the snail, swim onto blades of grass and form resistant cysts called metacercariae. These are ingested by the host animal, form immature flukes in the intestine and migrate to the liver where they mature into adults. As the flukes pass through the liver they feed on it until they come to a bile duct running through the liver, where they tend to congregate. The adults produce eggs in the bile ducts which pass out in the faeces via the gall bladder and intestines.

The life cycle of the fluke is quite long in comparison to the other internal parasites described above, twenty weeks or more, with the weather influencing both the hatching of the eggs and the activity of the snails. This means that symptoms of the flukes picked up in the summer will not be seen until December or January.

The liver fluke can over-winter in a number of ways, firstly as an adult fluke in the

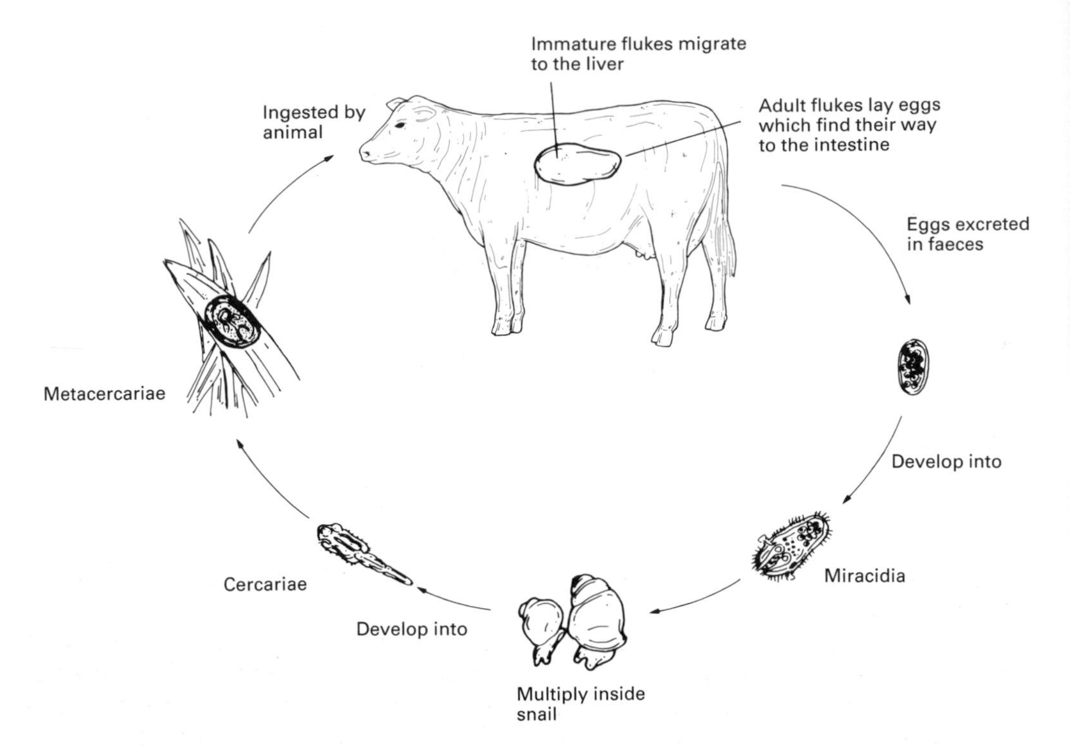

FIG. 23.3 The life cycle of the liver fluke.

liver, secondly as resistant metacercariae on the grass and, thirdly, inside dormant snails.

Symptoms

Sheep suffer from two types of disease—acute and chronic, whilst cattle normally only suffer the chronic disease.

Acute liver fluke in sheep is caused by the animal ingesting huge numbers of larvae. As these begin to migrate through the liver, they cause immense damage resulting in anaemia and sudden death.

The symptoms of the more common chronic liver fluke damage in both cattle and sheep are a progressive loss of body condition, anaemia and a collection of fluid under the jaw known as 'bottle-jaw'.

Prevention and control

Anthelmintics are available that will kill the fluke at most stages of its life cycle whilst in the primary host. These should be used under the direction of a veterinary surgeon as only a limited number are safe to use on milking cows.

The secondary host, the snail, lives in wet areas. Draining or fencing off these areas will help reduce the fluke population.

A fluke forecast is issued by the Ministry of Agriculture. This is determined by analysing the weather conditions that influence the hatching of the larvae.

Chapter 24

Zoonoses

Zoonoses (singular zoonosis) is the term used for diseases that are naturally transmissible between animals (including birds) and man.

Throughout the world there are well over one hundred such diseases, some causing minor irritation, others causing death. This chapter is concerned with those diseases that can be transmitted from farm animals—cattle, pigs, sheep—and that normally occur in the United Kingdom. For the sake of completing the picture, diseases transmitted from cats and dogs are also included.

Some of the diseases listed are, in terms of human health, notifiable (e.g. salmonella) and appear on the statute books under various Zoonoses Orders. In general, these Orders require any incidence of the listed diseases that occur in humans to be reported to the appropriate government office. The specifics of this legislation are not dealt with in this book.

The agricultural connection

Whilst farmers, stockpeople, veterinary surgeons and others closely connected with the agricultural industry are most at risk to a range of zoonotic infections, the ones that most affect the general public are food poisonings caused by such organisms as salmonella, listeria and campylobacter. These infections can cause death in the very young and elderly and are the cause of thousands of lost working days. Many food poisoning cases are a result of poor kitchen hygiene and poor food handling practices, but it must be remembered that foodstuffs, for example vegetables, can become contaminated on the farm of origin, so strict hygiene when handling any food for human consumption is essential.

Visitors to the farm are also at risk. In this age of 'diversification' there are many more farm 'open' days than there used to be, with school children and the general public being encouraged to visit farms and, in some cases, handle farm animals. Whilst this openness is to be encouraged, farmers should be aware that there is a small risk of a zoonotic infection being passed to visitors. Sensible precautions, such as those described at the end of this chapter, should significantly reduce this risk.

TABLE 24.1 *Zoonoses occurring in the United Kingdom.*

Bacterial	Parasitic	Other causal organisms
Anthrax	Cysticercosis	Chlamydia (psittacosis)
Brucellosis	Liver fluke	Orf
Campylobacteriosis	Hydatid disease	Q Fever
*Erysipelas	Toxocariasis	Ringworm
Leptospirosis	Toxoplasmosis	
Listeriosis		
*Pasteurellosis		
Salmonellosis		
**Streptobacillus*		
Streptococcus suis (meningitis)		
*Tuberculosis		
**Yersinia*		

*Not referred to in the text.

Zoonoses

There are twenty-one zoonoses referred to in the United Kingdom (see Table 24.1). The following section, divided into their causal organisms, describes the diseases that are either serious or that frequently occur.

The descriptions of modes of transmission and the symptoms that occur in humans are by no means exhaustive. Anyone suspecting a case of a zoonotic infection should consult his/her doctor.

Zoonoses caused by bacteria

Anthrax

This disease, also called 'wool-sorters' disease, is mainly trade related. That is, the people most at risk are those who handle skins, bones, blood and wool for a living. It is caused by the bacterium *Bacillus anthraci* whose spores can survive in the soil for many years. Humans are infected by direct contact with infected animals and animal products. The disease may enter the body via cuts in the skin, through the mouth/gut or inhaled into the lungs. Symptoms vary with the mode of entry:

via the skin—blackened painless skin lesions combined with a headache and high temperature. Meningitis may follow.
via the mouth/gut—acute stomach upset with bloody diarrhoea.
via the lungs—severe pneumonia. Usually fatal.

Prevention:

A vaccine is available for those in high risk occupations.
Wear protective clothing.
Ensure good ventilation of buildings.
Treat any wounds promptly.

Leptospirosis

This bacterial disease has two strains that are of particular concern to farmers and stockpeople:

Leptospira interrogans serovar. *hardjo* which causes 'dairy worker fever'.
Leptospira interrogans serovar. *icterohaemorrhagica* which causes Weil's disease.

L. hardjo is passed on via the dung and urine of infected animals, usually cows. Milking personnel who get splashed with dung and urine are most at risk. The human symptoms are an influenza-like illness that clears up after several days.

L. icterohaemorrhagica is passed on via the urine of rats. It used to be a disease of sewer workers, but these days water sports enthusiasts and others who come into contact with infected water are at risk. Weil's disease is characterized by jaundice and kidney failure (which does not occur with *L. hardjo*). The fatality rate from Weil's disease is up to 20 per cent.

Prevention:

Control rodents.
Drain wet ground.
Avoid swimming in or drinking contaminated water.
Cover cuts to the skin if in contact with contaminated water.
Early diagnosis if infection is suspected.

Campylobacteriosis

This disease is caused by the organism *Campylobacterfetus jejuni* which is excreted from most animals and birds. It causes food poisoning in humans.

The bacteria are killed by cooking and pasteurization, so outbreaks are normally due to poor kitchen hygiene, although water-borne and milk-borne outbreaks have occurred in the past.

Prevention:

Chlorinate drinking water.
Thoroughly cook meat, especially poultry.
Practise good kitchen hygiene.

Salmonellosis

There are over 1500 types of salmonella bacteria, with many that cause symptoms similar to food poisoning symptoms in humans. Most cases involve cattle and poultry, with only a few linked to sheep and pigs. The severity of the symptoms depends on the dose of infection, but generally a watery diarrhoea may last for about ten days with a possible risk of dehydration.

Prevention:

Practise good kitchen hygiene.
Cook foodstuffs thoroughly.
Avoid cross-contamination in the food store.
Good personal hygiene.

Listeriosis

The causal organism *Listeria monocytogenes* is found in many animals, birds, humans and in soil. It is also excreted in animal faeces. The symptoms of the infection include headache, high temperature and vomiting, with the possibility of meningitis. Babies and the elderly are at risk as well as pregnant women as it can cause abortion in late pregnancy.

The bacterium seems to be able to tolerate low temperatures and is able to survive in partially or poorly frozen foods. Cheese made from unpasteurized milk has been isolated as a high-risk food.

Prevention:

Store food at correct temperature.
Pasteurize milk.
Ensure that frozen food is heated to the correct temperature all the way through.

Streptococcal meningitis

Whilst there are a number of streptococci bacteria that may cause meningitis in humans, the main causal organism that farmers and stockpeople have to worry about is *Streptococcus suis*. Pigs carry the organism in their tonsils and noses. Infection is brought about by direct contact with infected animals or meat.

Prevention:

Take care when handling pigs.
Cover cuts to the skin when handling pigs.
Early diagnosis is essential.

Tuberculosis and brucellosis

In conclusion to this section on bacterial zoonoses it is worth mentioning bovine tuberculosis (TB) and brucellosis. National eradication schemes for cattle have all but eliminated these as human health problems. However, TB infection has recently been isolated in herds of deer with at least one case to date of the stockperson being infected.

Zoonoses caused by parasites

Toxocariasis

This is a common roundworm infection of dogs (*Toxocara canis*) and cats (*T. cati*) that can be passed on to children. The larvae develop into adult worms in the gut of dogs, puppies and cats, and are excreted in the faeces. Humans pick them up from contaminated soil and grass. The disease occurs mainly in children. The symptoms in children aged one to four years include asthma and bronchiole problems, sickness and vomiting. Older children aged between seven and eight can develop eye problems with progressive loss of vision leading to blindness.

Prevention:

Teach children good hygiene.
Keep dogs off play areas.
Cover children's sand pits when not in use.
Clean faeces from public places.
Worm dogs regularly.

Toxoplasmosis

This condition is caused by the protozoal parasite *Toxoplasma gondii*. Cats are the primary host, so it can be picked up from their litter, but it is also found in raw or undercooked meat, unwashed salads and unpasteurized goat's milk. Toxoplasmosis infection causes mild symptoms of fever, headache, loss of energy and possibly a cough. If pregnant women are infected however, it can be passed on to the fetus and may cause brain damage. In France, where undercooked meat is common in some dishes, it is estimated that one in every 2000 pregnancies is affected by *Toxoplasma*.

Prevention

Pregnant women should avoid cat litter.
Wear gloves when cleaning out cat litter.
Wash hands.
Cook meat thoroughly.

Hydatid disease

This disease is caused by the cysts of the tapeworm *Echinococcus granulosus*, the adult worms being found in dogs and foxes. The life cycle of the worm involves sheep as the main intermediate host, so people in contact with dogs from sheep farms are most at risk.

Symptoms vary with the site of the cyst, but one of the commonest forms, the liver cyst, causes stomach pains and jaundice.

Prevention:

Worm dogs regularly.
Stop dogs eating raw carcasses.
Wash vegetables.
Wash hands.

Other

Other, not so common, zoonoses caused by parasites include liver fluke and cysticercosis. Liver fluke may cause problems in bad fluke years and is usually traced back to wild watercress or contaminated watercress beds. Cysticercosis is caused by the larval stage of the tapeworm *Taenia saginata*. The infection is caused by eating raw and undercooked beef or pork. All slaughterhouses are required to inspect carcasses by cutting into selected areas in order to detect larval cysts. The carcass may be condemned if cysts are found.

Zoonoses caused by other agents

Chlamydia

This group of organisms causes enzootic abortion in sheep, but the disease of parrots and other birds, psittacosis, also comes under this heading. The organism is found in aborted material from ewes and in dust from the faeces and feathers of birds, mainly parrots, but also ducks, turkeys and pigeons. Shepherdesses are particularly at risk as the organism, picked up from lambing ewes, can cause late abortion in pregnant women.

Prevention:

Pregnant women should avoid ewes at lambing time.

Q fever

This is caused by *Coxiella burnetii*, which belongs to the rickettsia group of organisms. The infection is contracted through assisting births of cattle, sheep and goats. The organisms can live for months in animal litter and dust which may cause infection if inhaled. The symptoms include severe headache, high temperature and a cough, although recovery is rapid and the disease often goes undiagnosed. There may be heart complications in a few cases.

Prevention:

Wear gloves when assisting at births.
Avoid inhaling dust.

Orf

This is a viral skin infection contracted from infected sheep and lambs. The virus enters the skin via a cut or graze and causes a painful red lesion that may last three to six weeks.

Prevention:

Wash hands carefully after handling infected sheep.
Cover cuts to the skin when handling sheep.

Ringworm

This is a fungal skin infection contracted from cattle and horses, although a very similar fungus infects dogs and also causes ringworm.

The symptoms are scaly, crusty skin lesions that can affect any part of the body. If the scalp is affected, the result is temporary hair loss.

Prevention:

Avoid handling infected animals.
Wash hands after handling infected animals.

Personal preventive measures

Most of these measures are applied common sense and have been referred to in the prevention of the specific diseases. In conclusion however, they are worth reinforcing.

1. Wash hands after handling animals and before eating, drinking or smoking.
2. Gloves should be worn when handling aborted fetuses and afterbirths of all species. Fetuses should be disposed of by burial or burning so that dogs, cats and foxes cannot get access to them.
3. Pregnant women should avoid all contact with sheep and lambs at lambing time.
4. Ensure all pets are healthy; if in doubt, consult a veterinary surgeon. Worm dogs and cats regularly.
5. Most foodborne zoonoses can be avoided by hygienic kitchen practices, and careful handling and storage of food.
6. Any farm animal suspected of having a zoonotic disease should be treated as though it has until proved otherwise. Seek veterinary advice as soon as possible.

Chapter 25

Poisons and poisonous plants

The complexity of modern agriculture exposes farm animals to many poisons, with most poisoning cases resulting from animals gaining access to paints, sheep dips, insecticides and weed killers.

The symptoms resulting from poisoning may be vague, with abdominal pain and general dullness being the most common. Allied to this, it is quite difficult for a laboratory to examine a tissue sample and come up with a definite result; the laboratory usually requires a clue to the possible cause as a starting point. Trying to analyse a sample for every type of poison is a daunting task, so evidence from the farm, the veterinary surgeon and maybe post-mortem findings could suggest a possible poison. The size of the dose that contaminates the animal is also important. Some substances, like cyanide, are poisonous in very small quantities, but cyanide is sometimes used in minute quantities as a stimulant to initiate breathing in new-born animals.

Most substances taken into the animal body are quickly broken down or passed out of the system. Others, like some organochlorines, even though taken in small quantities tend to accumulate in the body. They are called cumulative poisons and often give rise to symptoms only after a period of time. In contrast to this, animals may develop a tolerance for a poison regularly taken in very small quantities.

Readers should note that the symptoms of poisoning vary dramatically according to circumstances and that those listed here are general examples. If poisoning cases are suspected, a veterinary surgeon should be consulted.

Mineral and compound poisons

Lead

Lead is probably the most common cause of poisoning in farm animals.
Whilst pigs seem to be able to cope with large quantities, calves and cows have relatively little tolerance. Their inquisitive nature makes the licking of unusual objects quite commonplace, often with drastic results.

Common sources of lead are:

Old paint.
Lead linings to old stable doors.
Contamination of pasture from petrol fumes and, in certain areas, from the waste tips of old lead workings.

The symptoms of the disease vary according to whether it is acute or chronic poisoning, with acute being most common. Affected animals may be blind and extremely excited. Cattle may bellow and try to climb up walls.

If lead poisoning is suspected you should consult your veterinary surgeon as it is a serious condition often leading to death.

Treatment is usually by intravenous injection of calcium disodium versenate which converts the lead that has been absorbed in the blood stream to an inert form that can be excreted from the body.

Copper

Copper is an essential element for growth, with copper deficiency being quite common. As copper is a cumulative poison however, there is a fine line between an adequate intake and an amount that will poison stock. Sheep are very sensitive to copper intake, cattle a little less so, whilst pigs can tolerate quite high levels. Indeed, copper is added in small quantities to pig fattening rations where it acts as a growth promoter.

Copper poisoning occurs where the pasture gets contaminated with copper either from spraying fruit trees or some other means, when copper is added to rations in the wrong quantities or when sheep get access to rations that contain copper, for example pig fattening rations. Indeed, the spreading of pig slurry on pastures has led to cases of chronic copper poisoning.

Symptoms include diarrhoea, abdominal pain and, as copper affects the liver, jaundice.

Creosote/diesel/paraffin

Animals will tend to lick diesel tanks and pipes and it is quite common to see cattle licking the engine of a tractor. Pigs will eat wood that has been painted with fresh creosote. These substances contain chemicals that will cause hyperkeratosis, a condition that shows up as hardening of the skin. This results in the loss of hair on the neck and shoulders, stunted growth and a discharge from the eyes.

Treating animals with vitamin A seems to help recovery.

Organochlorines/organophosphorus compounds

Both these compounds are used as insecticides. However, organochlorines in the form of BHC (louse powder), DDT and dieldrin (once used in sheep dips) are far less common than they used to be. Poisoning is as a result of a gradual build-up of poison in the fat of the animal.

Organophosphorus compounds are more common. It is a constituent of 'pour on' warble dressings, some anthelmintics, spray on fly repellents and is contained in sheep dips. (There are now sheep dips available that contain no organophosphorus compounds and so disposal of the waste water is environmentally safe.)

Organophosphorus compounds are quite toxic and are absorbed through the skin. Their toxicity may last for several days, so it could be a while after exposure to the chemical that the symptoms begin to appear. These include frothing at the mouth, possible blindness, diarrhoea and breathing difficulties.

Nitrates

During periods of drought or after heavy applications of manure, slurry or artificial fertilizers, the nitrate level in plants may build up to dangerous levels. In cattle, the rumen converts these nitrates to nitrites which are then absorbed into the blood stream where they combine with haemoglobin to form methaemoglobin. This compound is unable to release oxygen to the body tissues and the animal may die.

Pigs have been poisoned by eating swill that contained sodium nitrate, a compound used for curing meat. Death can occur quite quickly with the blood of affected animals being quite dark.

Cases of nitrate poisoning have occurred when cattle have gained access to fertilizer bags, bitten them open and licked at the contents.

Slug bait, rat bait and strychnine

These baits are used to control farm pests, strychnine being used to control moles.

Slug bait contains the chemical metaldehyde which, unfortunately, is made attractive to farm animals by the inclusion of cereals. A common rat bait used on farms contains warfarin. When injected, this stops the action of vitamin K which in turn interferes with blood clotting. Rat bait should only be placed in known or suspected rat runs and must be covered to prevent accidental poisoning to animals, especially pigs. Slug bait, however, is usually scattered over affected areas and every effort must be made to ensure that animals cannot gain access to treated areas.

Some symptoms are common for both types of poisoning, i.e. staggering and inactivity, but poisoning by warfarin may be suspected if there is blood in the dung.

The symptoms of strychnine poisoning are severe muscle spasms which can resemble the disease tetanus (*see* page 162).

Salt (sodium chloride)

Salt poisoning can occur in one of two ways:

When the animal has access to drinking water that contains high levels of salt, e.g. well or spring water.

When the animal is on dry food and its water intake is restricted.

Acute salt poisoning of cattle and pigs shows as abdominal pain and diarrhoea followed by nervous signs and blindness. Removal of the contaminated water and frequent, but limited, access to fresh water will relieve the symptoms if the animal is not too ill. If it is unable to drink, the water may have to be administered by stomach tube.

Chronic salt poisoning of pigs is fairly common and occurs after two or three days of restricted water intake. This may be caused by a blockage in the water system or drinking bowl or simply because the pigs are unable to reach the drinker. The symptoms are blindness, bumping into the wall and circling the pen. This is followed by convulsions and chomping of the jaw. Limited access to fresh water will relieve the symptoms if the pig is not too ill to take the water.

Antibiotics

Antibiotic poisoning can occur in cattle and sheep when they gain access to, or are accidentally fed, food contaminated with antibiotics. This will kill the normal bacteria in the rumen, the animal will go off its food and it may show a decrease in milk production.

Poisonous plants

Yew

The common yew (*Taxus baccata*) is frequently responsible for poisonings, although all varieties of yew grown in Britain are poisonous. The leaves, berries and bark are all poisonous, with cattle, sheep, pigs, goats and deer all being susceptible.

Yew contains the poison taxine which stops the heart. It kills very quickly and little can be done to help animals that have eaten it, although it is reported that black coffee may help.

Problems occur when animals break out of a field and gain access to the trees or when trees are blown over and the branches get scattered after high winds.

Laburnum

Laburnum is a dangerous plant with the bark, flower, leaves and, especially, the seeds in their pods all being poisonous. The poison is cystine, which causes excitement, uncoordinated movements and finally death. There is no specific treatment, although boiled black tea or coffee is sometimes used as a drench.

Ragwort

This yellow-flowered, bitter-tasting plant affects mainly cattle and sheep. Sheep will start to eat it despite its taste when it begins to overtake the grass on the field. When fed in hay or silage it loses the bitter taste and is quite acceptable to cattle. It causes severe liver damage which may not show up for several months. The symptoms include abdominal pain with straining that may induce a prolapse of the rectum.

There is no treatment and, as it may take several days for an animal to die, it may be best to have it slaughtered. As ragwort grows best on marginal land, the application of fertilizers and cultivations will help to control the plant.

Rhododendron

This is not a very common cause of poisoning, but it may occasionally affect cattle and sheep. The symptoms include vomiting (a rare occurrence in cattle and sheep) and excessive foaming and salivating from the mouth.

Kale

Kale can cause problems with growth and blood formation and needs to be considered in three separate areas:

1. Large intakes of kale over a short period of time can lead to a breakdown of red blood cells resulting in anaemia and blood in the urine. Frosted kale seems to be more dangerous.
2. Prolonged periods of low intake of kale can lead to anaemia caused by an upset in the blood formation mechanism.
3. Kale may interfere with the functioning of the thyroid gland, leading to goitre.

Bracken

The green bracken fronds are quite bitter and will only be eaten by very hungry animals. However, dried bracken is quite palatable, especially if it is fed with hay.

Cattle and sheep poisoned by the fresh green plant show scouring and blood in the dung. If they are poisoned by the dried bracken, the symptoms are severe anaemia as the bracken affects the formation of various blood cells.

Pigs eating bracken may suffer from a deficiency of thiamine. It is said that bracken rhizomes (i.e. roots) contain more poison than the leaves, something that should be borne in mind when ploughing land for reclamation or turning pigs out onto bracken-infested land.

Oak (acorns)

Large amounts of green acorns and oak leaves can be poisonous to farm animals, with cattle being the most susceptible.

Initially the cattle are dull with no appetite which then progresses to severe black-stained diarrhoea owing to the presence of blood from inflammation of the intestines.

Affected animals should be given hay to eat. Drenching with liquid paraffin may help.

In fields where large amounts of acorns are dropped from the trees, pigs are occasionally used to clear them up as they are more resistant to acorn poisoning than cattle or sheep.

Further reading

The following books contain specialist information on many of the topics mentioned in this book and should be consulted if more in-depth knowledge is needed.

AGRICULTURE DEVELOPMENT ADVISORY SERVICE (1989) *Summary of the Law Relating to Farm Animal Welfare.*

BELL, J.C., PALMER, S.R. and PAYNE, J.M. (1988) *The Zoonoses.* Edward Arnold, London.

BLOOD, D.C., RADOSTITS, O.M. and HENDERSON, J.A. (1983) *Veterinary Medicine.* Baillière Tindall, London.

BLOWEY, R.W. (1985) *A Veterinary Book for Dairy Farmers.* Farming Press, Ipswich.

BRENT, G. (1986) *Housing the Pig.* Farming Press, Ipswich.

CLARKSON, M.J. and FAULL, W.B. (1985) *Notes for the Sheep Clinician.* Liverpool University Press.

FRANDSON, R.D. (1981) *Anatomy and Physiology of Farm Animals.* Lea and Febiger, Philadelphia.

FRASER, A. and STAMP, J.T. Revised: CUNNINGHAM, J.M.M. and STAMP, J.T. (1987) *Sheep Husbandry and Diseases.* Collins, London.

GERRARD, F. (1977) *Meat Technology.* Northwood Publications Ltd., London.

HUNTER, R.H.F. (1982) *Reproduction of Farm Animals.* Longman, London.

MINISTRY OF AGRICULTURE, FISHERIES AND FOOD (1983) *Codes of Recommendations for the Welfare of Livestock: Cattle; Sheep; Pigs.*

PETERS, A.R. and BALL, P.J.H. (1987) *Reproduction in Cattle.* Butterworth and Co., London.

SAINSBURY, D. (1983) *Animal Health. Health and Disease and Welfare of Farm Livestock.* Granada Publishing, London.

SAINSBURY, D. (1986) *Farm Animal Welfare.* Collins, London.

SAINSBURY, D. and SAINSBURY, P. (1988) *Livestock Housing and Health.* Baillière Tindall, London.

SANDYS-WINSCH, G. (1984) *Animal Law.* Shaw and Sons Ltd., London.

THOMAS, D.G.M. with BEYNON, D.G., HERBERT, T.G.G. and LLOYD-JONES, J. (1983) *Animal Husbandry.* Baillière Tindall, London.

THORNTON, K. (1988) *Outdoor Pig Production.* Farming Press, Ipswich.

TOUSSAINT RAVEN, E. (1985 English Version) *Cattle Footcare and Claw Trimming.* Farming Press, Ipswich.

WALTON, J.R. (1987) *A Handbook of Pig Diseases.* Liverpool University Press.

WEBSTER, J. (1987) *Understanding the Dairy Cow.* B.S.P. Professional Books.

WEST, G. (Editor) (1988) *Blacks Veterinary Dictionary*, 16th Edition. A. and C. Black, London.

WHITTEMORE, C.T. (1980) *Lactation of the Dairy Cow.* Longman, London.

Index